PLANT ANALYSIS
RESEARCH METHODS

Chief Editor

S.S. Narwal

Department of Agronomy,
CCS Haryana Agricultural University,
Hisar - 125 004, India

Editors

O.P. Sangwan
O.P. Dhankhar

Department of Soil Science
CCS Haryana Agricultural University,
Hisar - 125 004, India

SCIENTIFIC PUBLISHERS (INDIA)

P.O. BOX 91 JODHPUR

Published by :
Pawan Kumar
Scientific Publishers (India)
5-A, New Pali Road, P.O. Box 91
JODHPUR - 342 001 (INDIA)
E-mail: info@scientificpub.com
www.scientificpub.com

ISBN: 81-7233-474-5

Reprint from Research Methods in Plant Sciences:
Plant Analysis, Vol. 4 (2007)
ISBN: 81-7233-469-9

Laser typeset : Rajesh Ojha
Printed in India

PREFACE

Allelopathy is a new field of science, as the term 'Allelopathy' was coined by Prof. Hans Molisch, a German Plant Physiologist in 1937. Till now lot of allelopathy resech has been done in various fields of Agricultural and Plant Sciences. However, there is no compilation of various Research Methods used. Every scientist is conducting research in his own way. It is causing lot of problems to researchers working in underdeveloped/Third World Countries in small towns without Library facilities. Therefore, to make available the standard methods for conducting allelopathy research independently, this multi-volume book has been planned. Since allelopathy is multi-disciplinary area of research, hence, volumes have been planned for each discipline.

In all the conferences held since 1990's in India or in Foreign countries, a need has always been felt for a Manual of Allelopathy Research Methods. During II International Allelopathy Conference, New Delhi (September 6-8, 1994), it was decided that a Book **"Methodology of Allelopathy"** may be compiled. Prof. S.S. Narwal, India and Prof. G.R. Waller, Oklahoma State University, Stillwater, USA agreed to prepare this Book. But the Project was later shelved. Sine then Prof. Narwal has been strongly feeling to prepare such a Book. After 10-years, Prof. S.S. Narwal has planned this multi-volume Book **Research Methods in Plant Sciences : Allelopathy.** Three volumes (Volume 1. Soil Analysis, Volume 2. Plant Protection and Volume 3. Plant Pathogens) of this Book have been released during the **IV. International Allelopathy Conference, August 23-25, 2004 at Haryana Agricultural University, Hisar-125004, India.** Two volumes (Volume 4. Plant Analysis and volume 5. Plant Physiology) will be released in November, 2006. Three volumes (Volume 6, Cell Diagnostics, Volume 7. Chemistry Methods and Volume 8. Weed Studies) are under preparation.

The Book series **"Research Methods in Plant Science : Allelopathy"** aims to provide basic information about the various methods to Researchers so that he can conduct research independently. The methods have been described in a simple way just like "DO IT YOURSELF" Book. Now it is hoped that researchers can do research independently by following steps given in this Book series.

This book consists of 12 Chapters, describing the methods to analyse various nutrients in plants. The Book is devided into two Sections : General and Determination of Plant nutrients. **The Section I. General**, provides very elementary and basic information about the various equipments and apparatus used to determine plant nutrients and preparation of Reagents etc. Further, methods of collecting plant samples and their digestion have been described. In **Section II. Determination of Plant Nutrients,** 8 Chapters describes methods of determining various plant nutrients (Carbon, Nitrogen, Phospherus, Potassium, Sodium, Calcium, Magnesium, Sulphur, Micronutrients and Toxic metals).

This book provides complete information for Plant Analysis in simple and lucid language. The Figures/Illustrations have been given at appropriate places in text. It will prove very useful to undergraduate and post graduate students and teaching Faculty for Class Room and Laboratory experiments as well as for research. We also express thanks to Sh. Pawan Kumar and Sh. Rajesh Ojha, Scientific Publishers, Jodhpur, India to print this Book.

We invite valuable suggestions from students and researchers, to make further editions of this Book more useful and meaningful.

November 6, 2006

S. S. Narwal
O. P. Sangwan
O. P. Dhankhar

CONTENT

Research Methods in Plant Sciences : Allelopathy.
Volume 4. Plant Analysis
Eds. S.S. Narwal, O.P. Sangwan and O.P. Dhankhar
Scientific Publishers (India), Jodhpur (2007) (pp. vii-x)

ALLELOPATHY AND PLANT ANALYSIS

S.S. Narwal

Department of Agronomy,
CCS Haryana Agricultural University, Hisar-125004, India
E-mail : narwal_1947@yahoo.com

1. INTRODUCTION

Allelopathy refers to any process involving secondary metabolites produced by plants, microorganism, viruses, fungi that influence the growth and development of Agricultural and Biological systems. It has been established that allelopathy offers great potential (a) to increase agricultural production (food grains, vegetables, fruits, forestry), (b) to decrease harmful effects of modern agricultural practices [multiple cropping, leaching losses from N fertilizers, indiscriminate use of pesticides (weedicides, fungicides, insecticides, nematicides), tolerant/resistant biotypes in pest] on soil health/ productivity and on environment and (c) to maintain soil productivity and a pollution free environment for our future generations. It is likely that in the near future allelopathy will be used in crop production, crop protection, agroforestry and agrohorticultural practices in developed and developing countries. Allelopathy may become one of the strategic sciences to reduce the environmental pollution and to increase agricultural production in Sustainable Agriculture of the 21st Century. The Allelopathy provides basis to Sustainable Agriculture, hence, currently the Allelopathy research is being done in most Countries Worldwide and is now receiving more attention from agricultural and bioscientists.

Allelopathy is a new field of science, as the term 'Allelopathy' was coined by Prof. Hans Molisch, a German Plant Physiologist in

1937. Therefore, till now there is no Book on **Methodology of Allelopathy Research.** It is causing lot of problems to researchers working in underdeveloped/Third World Countries, in small towns without Library facilities. Therefore, to make available the standard methods for conducting allelopathy research independently, this multi-volume book has been planned. Since allelopathy is multi-disciplinary area of research, hence, volumes have been planned for each discipline.

2. ALLELOPATHY AND PLANT NUTRIENTS

The adequate supply of plant nutrients is necessary for the normal growth and development of plants. The excess or deficiency of plant nutrients adversely influence the various growth processes and production of alleloachemicals. In general, most of the soils are deficient in major plant nutrients and under such conditions Interference (Competition + Allelopathy) starts between the various plant spp. growing together. It becomes very intense as the plant enters grand growth phase. Then plants start interfering with the neighbouring plants, leading to their decreased growth. In natural grasslands, the deficiency of nitrogen leads to succession, where, new plant spp with less nitrogen requirements start growing, replacing the spp with large requirements of nitrogen. Under these conditions, plants start releasing allelochemicals in soil, which are toxic to neighbouring plants spp leading to their reduced growth and death, so that all nitrogen is available to its own plants (Rice, 1984).

Plant nutrients are closely associated with allelopathy, both in the synthesis of allelochemicals and influence of allelochemicals on various physiological processes in plants. These physiological processes effects the seed germination, seedling growth, development and senescence of plant organs and plant maturity. Besides, plant nutrients also influence the synthesis of allelochemicals in plants and their release into the environment. Various factors (a) radiation, (b) mineral deficiencies, (c) water stress, (d) temperature, (e) allelopathic agents, (f) age of plant organs, (g) genetics and (h) pathogens and predators affect the amount of allelochemicals produced in plants (Rice, 1984). All these factors influence various physiological processes, which affects the synthesis of allelochemicals. These allelochemicals after the release into the environment influences the plant growth and development through the changes of following physiological processes (Rice, 1984): (a) cell division and cell elongation, (b) phytohormone induced growth, (c) membrane permeability, (d) mineral uptake, (e) availability of soil phosphorus and potassium, (f) stomatal opening (g) photosynthesis, (h) respiration, (i) protein synthesis and changes in lipid and organic acid metabolism, (j) inhibition of porphyrin synthesis, (k) inhibition or stimulation of specific enzymes, (l) corking and clogging of xylem elements, stem conductance of water and internal water relations and (o) miscellaneous mechanisms.

REFERENCES

Narwal, S.S. (1994). *Allelopathy in Crop Production*. Scientific Publishers (India), Jodhpur, India pp. 288.

Rice, E.L. (1984). *Allelopathy*. II Edition. Academic Press, New York. pp. 422.

Section I.

GENERAL

Research Methods in Plant Sciences : Allelopathy.
Volume 4. Plant Analysis
Eds. S.S. Narwal, O.P. Sangwan and O.P. Dhankhar
Scientific Publishers (India), Jodhpur (2007) (pp. 1-18)

1

BASICS OF ANALYTICAL REAGENTS AND PREPARATION OF STANDARD SOLUTIONS

V.K. Phogat and O.P. Sangwan

Department of Soil Science, CCS Haryana Agricultural University, Hisar-125004, India

1. INTRODUCTION

The analysis of plant samples for quantitative estimation of nutrient elements involves a variety of chemicals (acids, bases and their salts), called reagents. The accuracy of analysis of results, depends on (i) purity of reagents and (ii) preparation of exact standards of required reagents. It is, therefore, necessary to know the basic properties of reagents, preparation and standardization of reagent solutions. This chapter provides in-laboratory-preparation and standardization of reagents required for plant or soil analysis.

2. ANALYTICAL REAGENTS

These are chemicals (acids, basis, salts and their derivatives) used in specific concentrations to estimate the elemental composition of plants. The purity of these chemicals determines the grade of analytical reagents. The commercially available reagents are classified in three grades depending on their purity (Jackson, 1967).

2.1. Grades of reagents

(i) *Analytical Reagents (AR):* These reagents are of highest purity and called AR. Their impurities are removed during manufacturing, hence, their cost is highest. These are used where most accurate and micro level analysis is needed. To analyse micro-nutrients in plants, AR grade reagents are required.

(ii) *Laboratory Reagents (LR):* These reagents are next in purity to AR and are relatively cheaper than AR. In most of laboratory analysis work, these reagents are used.

(iii) *Guaranteed Reagents (GR):* These reagents rank next to LR in purity and cost, and are used for general soil and plant analyses.

2.2. Concentrated acids and bases of pure grade

Some specific properties of commonly used concentrated acids and bases (Jackson, 1967, Gupta, 1999) are listed in Table 1. The dilute standard solutions of these reagents (N/10, N/100) are used in various titrimetric analytical methods to estimate nutrient elements in soil and plant. Therefore, the volume of concentrated acids/bases required to prepare 1N solution is also given (Table 1).

Table 1. Specific properties of important concentrated acids and bases

Sr. No.	Acid / Base	Concentration (% w/w)	Sp. Gravity at 20°C	Approx. Normality	Vol. (mL) required for 1L of 1N sol.
1.	Acetic acid (CH_3COOH)	99.8	1.06	17.5	58
2.	Ammonium hydroxide (NH_4OH)*	28.3	0.90	15.0	67
3.	Hydrochloric acid (HCl)	39.0	1.19	11.6	78
4.	Hydrofluoric acid (HF)	48.0	1.16	27.8	36
5.	Nitric acid (HNO_3)	71.0	1.42	16.0	62
6.	Phosphoric acid (H_3PO_4)	86.5	1.71	45.0	23
7.	Perchloric acid ($HClO_4$	70.0	1.67	11.6	86

* Saturated solution of ammonia

2.3. Common analytical reagent salts

Salts are produced when acids and bases react. Metal salts may be produced in reaction of acids. Some salts are simple, while, others are complex in their composition. When salts are dissolved in water, they get ionized and dissociated into their sub-units as positively and negatively charged particles known as cations and anions, each consisting of one or more elements. The elements of various salts, acids and bases used in soil and plant anlysis are listed in Table 2.

Table 2. Atomic number and atomic mass of some common elements

Element	Symbol	Atomic Number	Atomic Mass
Aluminum	Al	13	27
Arsenic	As	33	75
Barium	Ba	56	137
Boron	B	5	11
Bromine	Br	35	80
Cadmium	Cd	48	112
Calcium	Ca	20	40
Carbon	C	6	12
Chlorine	Cl	17	35.5
Chromium	Cr	24	52
Cobalt	Co	27	59
Copper	Cu	29	63.5
Fluorine	F	9	19
Hydrogen	H	1	1
Iodine	I	53	127
Iron/Ferrous	Fe	26	56
Lead/Plumbus	Pb	82	207
Lithium	Li	3	7
Magnesium	Mg	12	24
Manganese	Mn	25	55
Mercury	Hg	80	201
Molybdenum	Mo	42	96
Nickel	Ni	28	59
Nitrogen	N	7	14
Oxygen	O	8	16
Phosphorus	P	15	31
Platinum	Pt	78	195
Potassium/Kalium	K	19	39
Selenium	Se	34	79
Silicon	Si	14	28

Silver	Ag	47	108
Sodium/Natrium	Na	11	23
Sulphur	S	16	32
Tin/Stanus	Sn	50	119
Zinc	Zn	30	65

The salts used in soil and plant samples analysis and their molecular and equivalent weights are given in Table 3.

Table 3. Molecular and equivalent weights of some reagent salts used in soil and plant analysis

Sr. No.	Compound	Formula	Mol. Wt. (g)	Eq. Wt. (g)
1	Aluminum chloride	$AlCl_3$	133.34	44.45
2	Aluminum oxide	Al_2O_3	101.96	33.99
3	Ammonium acetate	CH_3COONH_4	77.08	77.08
4	Ammonium chloride	$NH_4\,Cl$	53.49	53.49
5	Ammonium ferrous sulphate	$(NH_4)_2\,SO_4\,FeSO_4.6H_2O$	392.13	196.06
6	Ammonium fluoride	$NH_4\,F$	37.04	37.04
7	Ammonium metavanadate	$(NH_4)_2VO_3$	116.98	58.49
8	Ammonium molybdate	$(NH_4)_6Mo_7O_{24}.4H_2O$	1235.86	205.97
9	Ammonium nitrate	$NH_4\,NO_3$	80.04	80.04
10	Ammonium sulphate	$(NH_4)_2SO_4$	132.14	66.07
11	Barium acetate	$(CH_3COO)_2\,Ba$	255.43	127.72
12	Barium chloride	$BaCl_2.2H_2O$	244.28	122.14
13	Boric acid	H_3BO_3	61.80	20.61
14	Calcium acetate	$(CH_3COO)_2\,Ca$	158.00	79.00
15	Calcium carbonate	$CaCO_3$	100.09	50.05
16	Calcium chloride	$CaCl_2.2H_2O$	147.02	73.51
17	Calcium hydroxide	$Ca(OH)_2$	74.00	37.00
18	Calcium nitrate	$Ca(NO_3)_2$	164.00	82.00
19	Calcium sulphate	$CaSO_4.2H_2O$	172.17	86.08
20	Cobaltus chloride	$CoCl_2$	129.84	74.92
21	Copper sulphate	$CuSO_4.5H_2O$	249.69	124.85
22	Ferrous sulphate (hydrated)	$FeSO_4.7H_2O$	278.00	139.00
23	Magnesium chloride	$MgCl_2.6H_2O$	203.30	101.65
24	Magnesium nitrate	$Mg(NO_3)_2.6H_2O$	256.41	128.20
25	Magnesium sulphate	$MgSO_4.7H_2O$	246.00	123.00
26	Manganese sulphate	$MnSO_4.H_2O$	169.02	84.51
27	Oxalic acid	$C_2H_2O_4$	90.20	45.10
28	Oxalic acid (dihydrated)	$C_2H_2O_4.2H_2O$	126.07	63.03
29	Potassium chloride	KCl	74.50	74.50

30	Potassium dichromate	$K_2Cr_2O_7$	294.00	49.00
31	Potassium dihydrogen phosphate	KH_2PO_4	136.00	136.00
32	Potassium hydrogen phthalate	$COOHC_6H_4COOK$	204.22	204.22
33	Potassium hydroxide	KOH	56.00	56.00
34	Potassium iodide	KI	166.01	166.01
35	Potassium nitrate	KNO_3	101.10	101.10
36	Potassium permangnate	$KMnO_4$	158.00	31.60
37	Potassium sulphate	K_2SO_4	174.27	87.13
38	Silver nitrate	$AgNO_3$	169.87	169.87
39	Sodium acetate	$CH_3COONa.3H_2O$	136.00	136.08
40	Sodium acetate (anhydrous)	CH_3COONa	82.04	82.04
41	Sodium bicarbonate	$NaHCO_3$	84.01	84.01
42	Sodium carbonate	Na_2CO_3	106.00	53.00
43	Sodium chloride	NaCl	58.45	58.45
44	Sodium fuoride	NaF	41.99	41.99
45	Sodium hydroxide	NaOH	40.00	40.00
46	Sodium nitrate	$NaNO_3$	84.99	84.99
47	Sodium oxalate	$Na_2C_2O_4$	134.00	67.00
48	Sodium sulphate	Na_2SO_4	142.04	71.02
49	Sodium thiosulphate	$Na_2S_2O_3.5H_2O$	248.18	248.18
50	Zinc sulphate	$ZnSO_4.7H_2O$	287.54	143.77

3. TERMINOLOGY

The common terms used in plant analysis are as under:

I. **Atom and atomic weight:** Atom is the smallest particle of an element, which retains its properties and participates in a chemical reaction. Atomic mass/weight is the average weight of atoms of an element in atomic mass units (amu) per atom. The atomic mass of various elements is given in Table 2. The atomic mass/weight is expressed in grams and is termed as gram atom. One gram atom of different elements will be in proportion to their amu values, e.g., for carbon it is 12, for oxygen 16 and for hydrogen 1.

II. **Molecule and molecular weight:** A molecule is the smallest part of an element or compound which exists in a free state and posses the original properties. For example, H_2 and O_2 are molecules of hydrogen and oxygen (elements) and H_2O is a molecule of water (compound). Molecular weight of a substance (element/compound) is defined as a weight of one molecule of a substance relative to the weight of the carbon atom taken as 12. The molecular weight of a substance is represented by the sum of the atomic weights of the elements present in one molecule of the substance, for example, the molecular weight of oxygen (O_2) is 16x2=32. Similarly, the molecular weight of NaCl is 23+35.5 = 58.5 and of $CaCl_2$ is 40 +2(35.5) = 40+71 = 111. When the molecular weight of a substance is expressed in grams, it is termed as 1g mole of substance. For example, 58.5 g of NaCl means 1 g mole of NaCl, 32g of O_2 means 1 g mole of oxygen. The molecular weights of some commonly used compounds are given in Table 2.

III. **Equivalent weight:** The equivalent weight of a substance is the weight of an element or compound which combines with (or displaces from combination) 3.00 parts by weight of carbon or 1.008 parts by weight of hydrogen or 8.00 parts by weight of oxygen. The equivalent weights of substances, for practical purposes, are expressed in grams and are termed as gram equivalents. The equivalent of some commonly used compounds is given in Table 2.

IV. **Titrimetry and titration:** Titrimetry is a technique/method of quantitative determination of a substance in a given solution by titration with a standard of known strength. Titration is defined as the process of determining the volume of a substance required to just complete the reaction with a known amount of other substance. The substance of accurately known strength used in titration is called titrant and the substance to be determined is called titrate.

V. **Standard solution:** A solution whose strength/concentration is known is called a standard solution. A standard solution contains a definite number of gram equivalents or gram moles or gram weight per unit volume of the solution. For example, if a solution of NaCl contains (23+35.5) = 58.5 g in one litre solution, the solution contains one gram equivalent per litre of the solution and hence it is a standard solution and its strength is 1N. Also, if 5 g of NaCl is dissolved in 100 ml of the solution, its concentration is 5% and hence, is a standard solution.

VI. **Normal Solution:** A solution which contains one gram equivalent of a reagent in one litre of solution is termed as a normal solution and is denoted as 1N. For instance, if one gram equivalent NaCl =23+35.5 = 58.5 grams is dissolved in adequate distilled water and the final volume is made to 1.0 litre, the solution so obtained is a 1N solution. Similarly, 1N solution of KCl can be prepared by dissolving (39+35.5) = 74.5 grams of KCl salt in adequate volume of water and then making the final volume of the solution upto 1.0 litre.

VII. **Melli-equivalent per litre solution:** A solution which contains 1/1000 g equivalent of a solute (substance) in one litre of the solution. For example, if 58.5 mg (i.e. 1/1000 of 58.5 g of NaCl) of NaCl is dissolved in about 300 ml distilled water and then final volume of 1.0 litre is made with distilled water.

VIII. **Molar solution:** A molar solution is the solution which contains one gram molecular weight of the solute dissolved in 1.0 litre of solution and is denoted by M. If one gram molecular weight of the reagent is dissolved in 1.0 litre of solution, the solution is 1M and if two gram molecular weight is dissolved in 1.0 litre of the solution, it is called 2M solution.

IX. **Molal solution:** A molal solution is the solution, which contains one gram molecular weight of the solute dissolved in 1000 g of the solvent (water). For example, if one gram molecular weight of NaCl = 23 + 35.5= 58.5 g is dissolved in 1000 g of water, the solution so obtained is a molal solution. For practical purposes, molal solutions are not ordinarily used for analytical works.

X. **Per cent solution (by weight):** The concentrations are express- ed in grams of the solute per 100 g of the solution. For exam- ple, a 5% solution of KNO_3 contains 5 g of KNO_3 in 95 g of water.

XI. **Per cent solution (by volume):** The concentrations are expressed in mL of the solute per 100 mL of the solvent. For example, a 25% solution of an alcohol (ethanol) contains 25 mL of alcohol (ethanol) in 75 mL of distilled water.

XII. **Parts per Million (ppm):** The contents are expressed in number of parts (weight or volume) per one million parts of the solution. For example, a 5 ppm solution of P contains 5 mg P per litre (1000 mL) of the solution.

XIII. **Buffer solutions:** These solutions are used as standard solutions for setting the pH meters in pH dependent titrations and other pH based analytical work. These solutions have definite pH which does not undergo any change with passage of time or small dilutions. Buffer solution contains a weak acid (CH_3COOH) and its salt with a strong base (NaOH) or a weak base (NH_4OH) and its salt with a strong acid (HCl). For example, a solution of 'CH_3COOH and CH_3COONa' or of 'NH_4OH and NH_4Cl' is a buffer solution.

XIV. **Indicator solutions:** Indicator solutions are the solutions of such substances which indicate the completion of a chemical reaction in titration by a change of colour. Indicators used co- mmonly in volumetric analysis are of various types as under:

Acid-base indicators (pH based indicators): These are the indicators which are used for titrations involving acids and bases. A pH based indicator is the solution containing a substance (organic chemical) which is either a weak acid or a weak base and gives one color when it exists in unionized form and a different color when it gets ionized during the titration as the end point is reached (reaction between acids and bases just complete). Some of such indicators used in soil and plant analysis as listed in Table 4.

Table 4. Acid-base indicators and their colours in acid and base media

Name of indicator	pH range	Colour in acid medium	Colour in alkaline medium
Methyl orange	3.1-4.5	Red	Yellow
Methyl red	4.2-6.3	Red	Yellow
Bromothymol blue	6.0-7.6	Yellow	Blue
Litmus	5.0-8.0	Red	Blue
Phenolphthalien	8.0-10.0	Colourless	Pink
Thymolphthalein	9.3-10.5	Colourless	Blue

The selection of an indicator will depend on the type of the titration i.e. whether the titration is between a strong acid and strong base or a strong acid and weak bases or any such other combination (Table 5).

Table 5. Suitability of pH dependent indicators for acids and bases titrations

Type of titration	Examples of reaction	Suitable indicator	pH range of colour change
Strong acid and strong base	NaOH vs HCl NaOH vs H_2SO_4 KOH vs HCl etc.	Methyl orange, methyl red, bromothymol blue, phenolphthalien	5-10
Strong acid and weak base	HCl vs Na_2CO_3 HCl vs $NaHCO_3$ HCl vs Ca $(OH)_2$ H_2SO_4 vs Na_2CO_3 H_2SO_4 vs NaOH etc	Methyl orange, methyl red, (Not phenolphthalien)	4-6
Weak acid and strong base	CH_3COOH vs NaOH CH_3COOH vs KOH	Phenolphthalien, thymol-phthalein (Not methyl orange, or methyl red)	8-10
Weak acid and weak base	CH_3COOH vs Na_2CO_3 CH_3COOH vs $NaHCO_3$	No suitable indicator (Titration can not be done by using any indicator but potentiometric titration can be done)	Slow change in pH

B. Precipitation Indicators: In titration of neutral solution of NaCl (or choloride ion) with $AgNO_3$ solution, K_2CrO_4 solution is used as an indicator. In this case, at the end point of reaction when entire Cl^- ions have been precipitated with silver nitrate, CrO_4^{2-} ions will immediately react with the next drop of $AgNO_3$ to give brick red colour to indicate the end point.

In other type of precipitation reaction, where Ag^+ is titrated in presence of free nitric acid with Na or K (NaCNS or KCNS) first produces a precipitate of silver thiocyanates. When AgSCN is complete, the slightest excess of thiocyanate produces a reddish brown colour, due to formation of complex ferri-thiocyanate ion. These indicators are used to determinate chlorides, bromides and iodides in acid solution.

C. Self indicator : When one of reacting substance itself acts as the indicator it is called a self indicator. A standard solution of $KMnO_4$ acts as a self indicator, whenever titatred with oxalic acid or ferrous sulphate in presence of sulphuric acid. The end point has its own pink colour, which on completion of reaction gets decoloured on the formation of colourless products such as $MnSO_4$, K_2SO_4 and H_2O.

4. PREPARATION AND STANDARIZATION OF REAGENT SOLUTIONS

The volumetric titration methods are commonly used for chemical analysis of soil and plant samples and use various types of standard solutions. The standard solutions can be prepared by dissolving a known weight of the solute in gram equivalents/moles or their equivalent fractions in unit volume (1.0 litre). To prepare a solution of 1 N strength, dissolve exactly one gram equivalent of the solute in adequate distilled water taken in a 1.0 litre volumetric flask and finally make the volume up to the mark. The solution so obtained will be 1 N provided the solute is a pure and unhygroscopic. But all substances (reagents/chemicals) may not be available in pure form and unhygroscopic in nature. Some of them are very hygroscopic, (e.g. NaOH) and absorb substantial moisture during weighing, while other may not have 100% purity. Therefore, there is always a chance that a solution prepared from such reagents may be of lower strength/ concentration than what has been intended. Hence, it is necessary that the strength of such reagent solutions is tested against the *primary standard solutions*. A primary standard solution is one, which can be prepared from the pure and unhygroscopic reagent by drying and weighing exact required quantity of solute per unit volume of solution and with which the strength of other solutions are standardized (Chopra and Kanwar, 1976; Puri and Sharma, 1981). The substances viz., sodium carbonate (Na_2CO_3), sodium oxalate

(COONa), borax ($Na_2 B_4O_7.10H_2O$) succinic acid ($(CH_2\text{-}COOH)_2$) etc. are common primary standards.

4.1. Solutions of acids

The commercial available concentrated acids (Table 2.2.1) have very high strengths (normality) and as such these acids can not be used for titrations. These are, therefore, diluted with distilled water. The extent of dilution depends upon the original strength of the acid and the strength required for a given titration.

4.1.1. *Preparation of 1 N solution of H_2SO_4*

Principle

An acid can be diluted using the equation $N_1V_1 = N_2V_2$, where V_1 is the volume of concentrated acid to be diluted with normality N_1 and V_2 is the volume of the standard acid to be prepared with the required normality of N_2.

Apparatus

Volumetric flask, conical flask, burette, pipette, measuring cylinder and beakers.

Reagent

(i). 36 N Sulphuric acid (H_2SO_4)

(ii). Distilled water

Procedure

(i) Calculate the volume of the concentrated H_2SO_4 required to be diluted to 1N using the principle equation as under:

$$N_1V_1= N_2 V_2$$

Where, N_1 = Normality of the concentrated acid (36 N)

V_1 = Volume of the concentrated acid to be calculated?

N_2 = Normality of the solution to be prepared (1N)

V_2 = Volume of the solution to be prepared (1.0 litre)

Now putting the values, in equation $N_1V_1 = N_2V_2$

$$V_1 = \frac{N_2V_2}{N_1} = \frac{1 \times 1000}{36} = 28 \text{ mL.}$$

(ii) Take 28 mL of concentrated acid in a measuring cylinder and transfer it to a 500 mL beaker containing 300 mL of distilled water.

(iii) Transfer the contents of the beaker in a 1.0 litre volumetric flask and make the volume with distilled water up to the mark. This will be 1N solution of H_2SO_4, which is standardized using a primary standard solution under:

4.1.2. *Standardization of the prepared 1N solution of H_2SO_4*

Principle

A solution of acid of unknown strength can be standardized by titrating it against a primary standard solution of a base using an appropriate indicator.

Apparatus

Volumetric flask, conical flask, burette, pipettes, measuring cylinder, China dish, analytical balance and beakers.

Reagent

(i) 1 N Sodium carbonate (Na_2CO_3)– primary standard solution

(ii) Methyl red indicator

(iii) Distilled water

Procedure

(i) First prepare 1N solution of sodium carbonate (Na_2CO_3) by dissolving 52.98 g of well dried salt in distilled water and making the final volume 1.0 litre.

(ii) Fill the burette with 1N Na_2CO_3 solution.

(iii) Pipette out 10 mL of 1 N H_2SO_4 to be standardized in a 100 mL conical flask and add 2-3 drops of methyl red indicator, which will give pink color.

(iv) Titrate the contents of the flask with 1N solution of Na_2CO_3 taken in a burette till the pink color changes to yellowish. Note down the volume of the solution used from the burette. Let it be X mL.

(v) Now using the principle equation, calculate the exact normality of the H_2SO_4 solution prepared and its required volume to be diluted to 1 litre (1000 mL) for making it of xactly of 1N concentration as under :

Calculations

Let the normality of the prepared H_2SO_4 solution be (N_2)	=	to be calculated ?
Volume of the prepared H_2SO_4 taken for titration (V_2)	=	10 mL
Volume of 1 N Na_2CO_3 used for titration of H_2SO_4 (V_1)	=	X mL (where X>10 mL)
	=	say 10.5 mL
Normality of Na_2CO_3 used for titration (N_1)	=	1 N

Since, $N_1V_1 = N_2V_2$; therefore, $N_2 = \frac{N_1V_1}{V_2}$

Now putting the values, $N_2 = \frac{1 \times 10.5}{10} = 1.05$ N

Hence the normality of the prepared H_2SO_4 = 1.05 N (instead of 1 N)

Therefore, to have its normality exactly 1 N, the dilution of the prepared solution will be done by calculating its required volume, according to the equation:

$$N_1V_1 = N_2\,V_2,$$

Where N_1= Normality of the acid to be prepared = 1 N

V_1 = Volume the acid to be prepared =1000mL

N_2 = Normality of prepared acid = 1.05 N

V_2 = Volume of the prepared acid required = ?

$$\text{Or, } V_2 = \frac{N_1V_1}{N_2}$$

Now putting their values,

$$V_2 = (1 \times 1000)/1.05 = 952 \text{ mL}$$

Therefore, to prepare 1 lire of IN H_2SO_4, the volume of 1.05 N acid required is: 952 mL.

So, take 952 mL of 1.05 N acid and dilute to 1 litre with distilled water.

Exercise 1. Prepare 1.0 litre (1000 mL) of 0.1N H_2SO_4 solution

Procedure

(i) First, prepare 1.0 litre of 1N H_2SO_4 and standardized it with a primary standard as described in the section 4.1.2 above.

(ii) Now calculate by using the equation, $N_1V_1 = N_2 V_2$, the volume of 1N H_2SO_4 required to be diluted to 1 litre for obtaining 0.1N H_2SO_4 as explained below:

$$N_1V_1 = N_2 V_2$$

Where, N_1 = 1 N, normality of the standard H_2SO_4

V_1 = Volume of the standard H_2SO_4 to be calculated?

N_2 = 0.1N, normality of the solution to be prepared

V_2 = 1000 mL, volume of the solution to be prepared

Putting the values in the equation: $V_1 = \frac{N_2V_2}{N_1}$

$$V_1 = \frac{0.1 \times 1000}{1} \text{mL} = 100 \text{ mL}$$

(iii) So, take 100 mL of 1N H_2SO_4 in a 1.0 litre volumetric flask and make the volume upto the mark with distilled water. The resulting solution is the required 1.0 litre solution of 0.1N H_2SO_4.

Exercise 2. Prepare 250 mL of 0.2N H_2SO_4 solution

Procedure

(i) First, prepare 1.0 litre of 1N H_2SO_4 and standardized it with a primary standard as described in the section 4.1.2 above.

(ii) Now calculate by using the equation, $N_1V_1 = N_2 V_2$, the volume of 1N H_2SO_4 required to be diluted to 250 mL for obtaining 0.1N H_2SO_4 as explained below:

$$N_1V_1 = N_2 V_2$$

Where, N_1 = 1N, normality of the standard H_2SO_4

V_1 = Volume of the standard H_2SO_4 to be calculated?

N_2 = 0.2N, normality of the solution to be prepared

V_2 = 250 mL, volume of the solution to be prepared

Putting the values in the equation: $V_1 = \frac{N_2V_2}{N_1}$

$$V_1 = \frac{0.2 \times 250}{1} \text{mL} = 50 \text{ mL}$$

(iii) So, take 50 mL of 1N H_2SO_4 in a 250 mL volumetric flask and make the volume upto the mark with distilled water. The resulting solution is required 250 mL solution of 0.2N H_2SO_4

4.2. Solutions of bases

Sodium hydroxide (NaOH) is a hygroscopic substance and is commonly contaminated with varying amounts of sodium carbonate due to the action of carbon dioxide of the air.

For preparing standard solution of NaOH pure pellets should be used.

4.2.1. *Preparation of 1N solution of NaOH*

Principle

A solution of 1N NaOH can be prepared by dissolving 1.0 g equivalent weight of well dried NaOH pellets in 1.0 litre of the solution.

Apparatus

Volumetric flask, conical flask, burette, pipette, analytical balance, wash bottle and beakers.

Reagent

(i). Sodium hydroxide (NaOH) pellets

(ii). Distilled water

Procedure

(i) Weigh 1.0 g equivalent of well dried NaOH pellets, i.e. 23+ 16 + 1 = 40 g of NaOH and dissolve it in about 300 mL of distilled water taken in a 500 mL beaker.

(ii) Transfer this solution to a 1.0 litre volumetric flask and make up the volume with distilled water up the mark. The resulting solution will be of approximately 1N strength.

4.2.2. *Standardization of the prepared 1N solution of NaOH*

Principle

The solution of a base of an unknown strength can be standardized by titrating it against a standard solution of an acid using an appropriate indicator.

Apparatus

Volumetric flask, conical flask, burette, pipettes and beakers

Reagent

(i). 1 N H_2SO_4 solution (already standardized with a primary standard solution)

(ii). Methyl red indicator

(iii). Distilled water

Procedure

1. Add 1-2 small pellets of NaOH in 1.0 litre of the solution prepared already (as described in section 4.2.1. above) so as to make its normality slightly higher than 1N.
2. Fill the burette with 1N H_2SO_4 solution.
3. Pipette out 10 mL of the NaOH solution to be standardized in a 100 mL conical flask and add 2-3 drops of methyl red indicator, which will give yellowish color.
4. Titrate the contents of the flask till the yellowish color changes to pink. Note down the volume of 1N H_2SO_4 used from the burette. Let it be X mL.
5. Now using the principle equation, calculate the exact normality of the NaOH solution prepared and its required volume to be diluted to 1 litre (1000 mL) for making it exactly of 1N strength as explained below.

Calculations

Let the normality of the prepared NaOH solution be (N_2)	= to be calculated ?
Volume of the prepared NaOH taken for titration (V_2)	= 10 mL
Volume of 1N H_2SO_4 used for titration of NaOH (V_1)	= X mL (where X >10 mL)
	= say 10.5 mL
Normality of H_2SO_4 used for titration (N_1)	= 1 N

Since, $N_1V_1 = N_2V_2$; therefore, the normality of NaOH prepared,

$$N_2 = \frac{N_1V_1}{V_2}$$

Now putting the values, $N_2 = \frac{1 \times 10.5}{10} = 1.05$ N

Hence the normality of the prepared NaOH = 1.05 N (instead of 1 N)

Therefore, to have its normality exactly 1N, the dilution of the prepared NaOH solution will be done by calculating its required volume to be diluted to 1 litre.

According to the equation:

$$N_1V_1 = N_2 V_2,$$

Where N_1= Normality of NaOH to be prepared = 1 N

V_1 = Volume the NaOH solution to be prepared =1 litre (1000mL)

N_2 = Normality of prepared NaOH solution= 1.05 N

V_2 = Volume of the prepared NaOH solution required = ?

$$\text{Or, } V_2 = \frac{N_1 V_1}{N_2}$$

Now putting their values,

$$V_2 = (1 \times 1000)/1.05 = 952 \text{ mL}$$

Therefore, to prepare 1 lire of IN NaOH, the volume of 1.05 N NaOH solution required will be 952 mL.

So, take 952 mL of 1.05 N NaOH solution and dilute to 1 litre with distilled water.

Exercise 1. Prepare 500 mL of 0.5N NaOH solution

Procedure

(i) First, prepare 1.0 litre of 1N NaOH and standardized it with a standard acid as described in the section 4.2.2 above.

(ii) Now calculate by using the equation, $N_1V_1 = N_2 V_2$, the volume of 1N NaOH required to be diluted to 500 mL for obtaining 0.5N NaOH as explained below:

$$N_1V_1 = N_2 V_2$$

Where, N_1 = 1N, normality of the standard NaOH

V_1 = Volume of 1N NaOH to be calculated?

N_2 = 0.5N, normality of the solution to be prepared

V_2 = 500 mL, volume of the solution to be prepared

Putting the values in the equation: $V_1 = \frac{N_2 V_2}{N_1}$

$$V_1 = \frac{0.5 \times 500}{1} \text{mL} = 250 \text{ mL}$$

(iii) So, take 250 mL of 1N NaOH in a 500 mL volumetric flask and make the volume upto the mark with distilled water. The resulting solution is the required 500 mL solution of 0.5N NaOH.

5. REFERENCES

Chopra, S.L. and Kanwar, J.S. (1976). *Analytical Agricultural Chemistry*. Kalyani Publishers, New Delhi, pp. 514.

Gupta, P.K. (1999). *Soil, Plant, Water and Fertilizer Analysis*. Agro-Botanica Publishers.

Jackson, M.L. (1967). *Soil Chemical Analysis*. Prentice Hall of India Pvt. Ltd., New Delhi, pp. 498.

Puri, B.R. and Sharma, L.R. (1981). *Principles of Physical Chemistry*. Vishal Publications, New Delhi, pp. 1026.

Research Methods in Plant Sciences : Allelopathy.
Volume 4. Plant Analysis
Eds. S.S. Narwal, O.P. Sangwan and O.P. Dhankhar
Scientific Publishers (India), Jodhpur (2007) (pp. 19-38)

2

ANALYTICAL INSTRUMENTS

V.K. Phogat, Rita Dahiya and O.P. Sangwan

Department of Soil Science, CCS Haryana Agricultural University, Hisar-125 004, India

1. INTRODUCTION

Analytical instruments are used to determine the elemental composition of plant tissues. Most of the instruments used are based on the interaction of electromagnetic radiations with the matter (colorimeter, spectrophotometer, flame photometer, atomic absorption spectrophotometer, x-ray diffraction etc). Owing to such interaction, either the energy is absorbed or emitted by the matter. These absorption or emission processes occur in the electromagnetic spectrum (gamma region to radio region). The methods to measure the emitted or absorbed radiation form experimental basis of different instruments (Willard *et al.*, 1965; Isaac and Kerber, 1971; Chatwal and Anand, 1995). Most of the instruments measure (i) the quantity of radiation energy absorbed or emitted by the material and (ii) the intensity of such radiation. The researchers who use Spectrophotometer, Flame Photometer, Atomic Absorption Spectrophotometer to analyse various nutrient in plant samples should know their principles, functioning and maintenance explained in this chapter.

2. ANALYTICAL INSTRUMENTS

The use of sophisticated instruments (spectrophotometer, flame photometer, atomic absorption spectrophotometer, etc.) is increasing in research due to high precision and greater efficiency than conventional titrimetric techniques or general instruments. Hence, these should be operated and maintained properly as under :

2.1. Spectrophotometer

2.1.1. *Introduction*

It measures the absorption or emission of radiations by the matter in the ultraviolet (380-200 nm) and visible (380-780 nm) regions of spectrum. If absorption or emission of radiations are measured in the visible region, it is called colorimeter. Spectrometer determines the elemental concentration in a given solution by measuring the relative absorbance or transmittance of visible radiations with respect to a known concentration of the element in a standard solution. The visible light is composed of seven colours of different characteristic wavelengths. If the energy of one or more wavelengths of visible light is absorbed greater than other wavelengths, the transmitted radiation will appear coloured instead of white. The colour of the solution is always the complement of the colour of incident/transmitted radiation. For example, a solution absorbing in the blue region will appear yellow and in red region will appear bluish green. The transmitted colours in the visible region, corresponding to various wavelengths and complementary apparent colour of the sample solution are given in Table 1.

Table 1. Relation between colour of transmitted light and sample solution

Wave length (nm)	Transmitted colour	Complementary colour of sample solution
380-435	Violet	Yellowish green
435-480	Blue	Yellow
480-490	Greenish blue	Orange
490-500	Bluish green	Red
500-560	Green	Purple
560-580	Yellowish green	Violet
580-595	Yellow	Blue
595-650	Orange	Greenish blue
650-780	Red	Bluish green

The spectrophotometer is commonly used to determine of the elements like N, P, S, etc., in soil and plant samples by using the specified methods, involving development of some color. The optimum setting of wavelength in the monochromator (wavelength selector) of the instrument for a given test solution can be done as per Table 1.

2.1.2. *Principle*

Spectrometer is based on the measurement of intensity of radiant energy absorbed in or transmitted through a sample solution. When light radiations are incident upon a homogeneous medium, a part of it is reflected, a part is absorbed by the medium and the remaining part is transmitted through the medium as such (Fig. 1).

If I_0 denotes intensity of incident light radiations, I_r the reflected, I_a the absorbed and I_t the transmitted light radiations, then I_0 can be written as:

$$I_0 = I_r + I_a + I_t \quad (1)$$

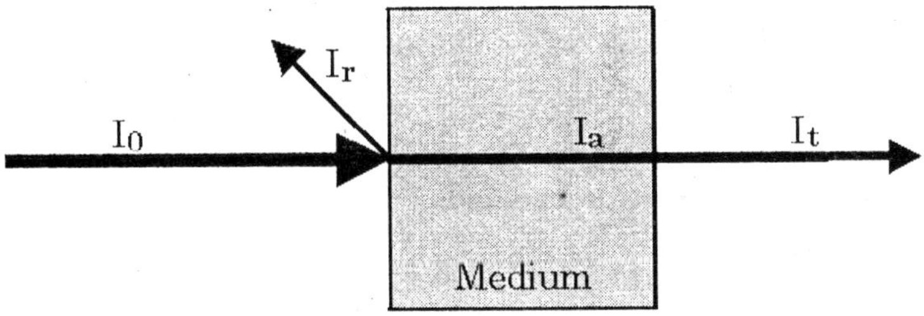

Fig. 1. Transmittance, absorbance or reflectance of a light radiation through a medium.

The I_r is negligible for air-glass interfaces, therefore, the equation (1) may be rewritten as:

$$I_0 = I_a + I_t \quad (2)$$

The intensity of I_a and/or I_t radiations through a medium of different thickness or concentration was described by Lambert's Law and Beer's Law, respectively.

Lambert's Law: When a beam of monochromatic light passes through a transparent homogeneous medium, the rate of decrease of intensity with the thickness of medium (x) is directly proportional to the intensity of incident radiations (I) and is expressed as:

$$-\left(\frac{dI}{dx}\right) \propto I \quad (3)$$

$$-\left(\frac{dI}{dx}\right) = kI \quad (4)$$

where k is the proportionality factor.

$$\text{or} \quad -\frac{dI}{I} = k\,dx \tag{5}$$

Integrating the above equation,

$$\ln I = -kx + A \tag{6}$$

where A is constant of integration.

Substituting $I = I_0$ at $x = 0$ in Eq. Or equation (6),

$$A = \ln I_0 \tag{7}$$

Combining equations 6 and 7

$$\ln I = -kx + \ln I_0 \tag{8}$$

$$\text{or } \ln\frac{I}{I_0} = -kx \text{ or } 2.303\log_{10}\frac{I}{I_0} = -kx \text{ or } \log_{10}\frac{I}{I_0} = -\frac{k}{2.303}x \tag{9}$$

$$\text{or} \quad \log_{10}\frac{I}{I_0} = -Kx \tag{10}$$

where K is absorption coefficient of the medium.

Taking antilogarithm on both sides of the above equation,

$$I = I_0 10^{-Kx}$$

Here I is transmitted light, so expressing it as I_t instead of I in the above equation

$$I_t = I_0 10^{-Kx} \text{ or } \frac{I_0}{I_t} = 10^{Kx} \tag{11}$$

Beer's law: When a beam of monochromatic light passes through a transparent homogeneous medium, the rate of decrease of intensity with the concentration of medium (c) is directly proportional to the intensity of incident radiations (I) and is expressed as:

$$-\frac{dI}{dc} \propto I \tag{12}$$

Analogous to Lambert's Law, the equation (12) is finally expressed as:

$$\frac{I_0}{I_t} = 10^{K'c} \tag{13}$$

On combining equations (11) and (13),

$$I_t = I_0 10^{-KK'cx} \quad \text{or} \quad I_t = I_0 10^{-acx} \quad \text{or} \quad \log\frac{I_0}{I_t} = acx \tag{14}$$

The equation (14) is mathematical statement of Beer-Lambert's law and is fundamental equation of colorimetry and spectrophotometry.

The $\frac{I_t}{I_0}$ is termed as transmittance (T) and the log $\frac{I_0}{I_t}$ is termed as absorbance (A) of the medium. The absorbance is also known as optical density or extinction coefficient.

$$A = \log\frac{I_0}{I_t} = \log\frac{1}{T} = D \qquad (15)$$

In equation (14), the value of 'a' depends upon the unit of concentration (c). If c is expressed in mole/L and thickness (x) in cm, then 'a' is replaced by the symbol ε and is called as molar absorption coefficient or molar absorptivity, and equation (14) becomes:

$$\log\frac{I_0}{I_t} = \varepsilon cx \text{ or } A = \varepsilon cx \qquad (16)$$

which indicates that optical density or absorbance is proportional to concentration (c), and thickness of sample (x). If a container for holding the sample solution, e.g. cuvette of thickness 1 cm is used then, equation (16) becomes:

$$A = \varepsilon c$$

According to this equation, absorbance is directly proportional to 'c'. If c = 1 mole/L, then A = ε.

As per Beer-Lambert's law, a straight line is obtained if optical density or absorbance is plotted against concentration on an ordinary graph paper.

2.1.3. *Components of Spectrophotometer/Colorimeter*

The basic components of spectrophotometers are : source of radiations, monochromator or filter, sample compartment, detector and read out system (Fig. 2).

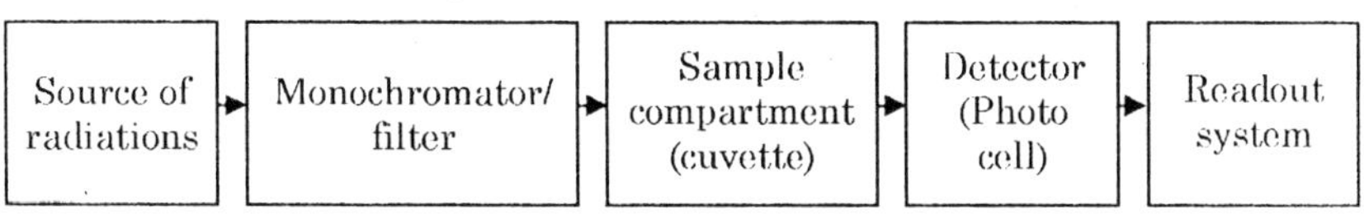

Fig. 2. Basic components of Spectrophotometer/Colorimeter

I. Source of radiations: A tungsten-filament lamp is used as source of radiations of visible light.

II. Monochromator or filter: Monochromator or filter is a device that selects required wavelength (from continuous spectra emitted by the radiation source) and allows them to pass through it. A monochromator has an entrance slit, a dispersing material (a prism or grating) and an exit slit. It instead of absorbing the undesired wavelengths, isolates the wide band of radiations with dispersing material and allows only the required wavelength to pass through the exit slit. The filter permits the light of the required wavelength to pass but absorbs the light of other wavelengths wholly or partially. A particular filter is used for a specific analysis, therefore, different filters are used for the analyses of different elements.

III. Sample compartment: It is a container (e.g., cuvette) for holding the sample for analysis and is transparent to the wavelength region being recorded. For visible regions, it is generally made of colour corrected fused glass. The internal thickness of a sample container is generally 1 cm.

IV. Detector: The detector is a photosensitive device to measure the intensity of radiations transmitted through the sample solution. Main photosensitive devices used are photovoltaic cells, phototubes and photomultiplier tubes.

V. Readout device: It gives out put reading for the intensity of radiations absorbed or transmitted through the sample. Now-a-days most of the readout systems give digital out put.

2.1.4. *Preparation of standard curve*

A standard curve for each element should always be prepared for a given spectrophotometer. To prepare standard curve for the element to be analysed, a series of solutions of known concentrations (standard solutions) of that element are prepared from the stock solution, which is prepared from a suitable salt. The colour in each standard solution is developed and its intensity in terms of absorbance or transmittance is measured on the instrument. First, the spectrophotometer is set at the desired wavelength. The absorbance of each solution is then measured and plotted against the concentrations of the solution to get a calibration curve i.e. straight line graph (Fig. 3). In practice, the standard solutions of such concentrations are prepared which cover the best absorbance range, say 0 to 0.6 absorbance units. If the concentrations are much higher than this range, the Beer's Law is not followed exactly and the accuracy of determination will be affected. Therefore, necessary dilutions should be made to bring the concentration within the range where accuracy is not affected.

The relation of transmittance to concentration is also shown in Fig. 3. When a beam of light is passed through distilled water, 100% of the light is transmitted and no light is absorbed; whereas, in the case of coloured solution, some light is transmitted while some of it is absorbed. Mathematically, absorbance (A) is related to logarithmic of inverse of transmittance (T) as given in equation (15) i.e. $A = \log \frac{1}{T}$.

The plotting of transmittance versus concentration of the test element on linear scale, results in a curve (Fig. 3) To get a straight line relationship, the transmittance is either plotted on logarithmic scale (Fig. 4) or the logarithmic of transmittance is plotted on linear scale against concentrations of test element.

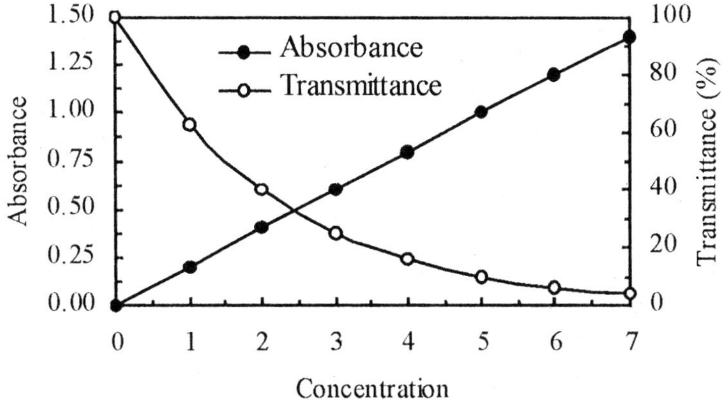

Fig. 3. Calibration curve of absorbance and transmittance with concentration

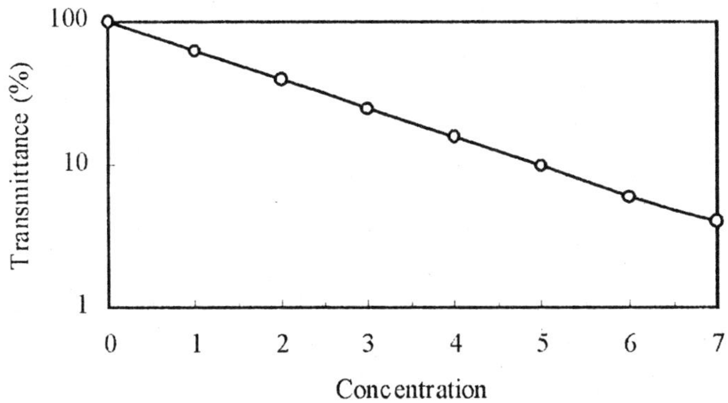

Fig. 4. Plot of transmittance on logarithmic scale against concentration of test element on linear scale to get a linear curve

2.1.5. *Operating instructions for Spectrophotometer/ Colorimeter*

(i) Switch on the instrument and allow it to stabilize (warm up) for about 15-20 minutes.

(ii) Select the desired wavelength, depending on the colour of the test solution.

(iii) Select the absorbance or transmittance mode as per the requirement.

(iv) Take the blank sample in a cuvette and close the lid. Press the read button and calibrate the instrument to zero absorbance or 100 % transmittance.

(v) After calibration, take the test sample in the cuvette, close the lid and note down the absorbance or transmittance reading. Similarly, record the readings of subsequent test samples.

(vi) Calibrate the instrument after every 8-10 samples.

(vii) Calculate the elemental concentration of the test sample by comparing the readings with the standard curve.

2.1.6. *Determination of element concentration in a test sample*

To measure the concentration of a test solution, (i) first the colour is developed by the prescribed method, (ii) then its absorbance is measured at a wavelength depending on colour (as explained for the calibration curve in section 2.1.4). (iii) calculate the concentration of test solution from the calibration curve by using the value of concentration against and the absorbance of the sample solution.

2.2. Flame Photometer

2.2.1. *Introduction*

In flame photometer, flame is used to provide energy for excitation of the atoms of an element introduced into the flame. It is a simple, rapid and sensitive instrument to determine the concentrations of elements (soil and plant samples) that can easily be excited, e.g., alkali metals (Na and K) and alkaline earth metals (Ca and Mg). When a liquid sample is introduced into a flame, it measures the colour and intensity of the light (radiation of characteristic wavelength) emitted by the excited atoms when they return to ground state. The colour of the emitted light determines the concentration of the element present in the sample.

2.1.2. *Principle*

When a liquid sample containing metallic salt is sprayed into a flame, the liquid gets vapourize leaving the particles of the solid salt. The salt is then converted into the gaseous state and a part or all of the gaseous molecules are progressively dissociated into free neutral atoms. These neutral atoms are excited by the thermal energy of the flame. The excited atoms, which are unstable, emit characteristic radiations and return to lower energy state and eventually reaching the unexcited/ground state. The emitted characteristic radiations are then measured.

The energy of radiation, E (ergs) emitted by an atom during jump from higher energy level, E_2 (excited state) to lower energy level, E_1 (ground state) is defined as:

$$E = E_2 - E_1 = h\nu \qquad (1)$$

where h is the Planks' constant with a value of 6.624 x 10^{-27} ergs sec and ν is the frequency of the emitted radiation, sec^{-1}. The ν is related to wave length, λ (cm) and velocity of light in vacuum, c ($3x10^{10}$ cm sec^{-1}) as:

$$\nu = \frac{c}{\lambda} \qquad (2)$$

Combination of equations (1) and (2) gives:

$$E = \frac{hc}{\lambda} \quad \text{Or} \quad \lambda = \frac{hc}{E} \qquad (3)$$

Using equation (3), we can calculate the wavelength of the emitted radiation, which is characteristic of the atom of a particular element. When the flame photometry is employed as an analytical tool, the wavelength of the radiation coming from a flame tells about the elements which are present in the sample and the intensity of the radiation enables us to know the amounts of the elements present.

The fraction of free atoms that are thermally excited is governed by a Boltzmann distribution which is as follows:

$$\frac{N}{N_0} = Ae^{-\frac{\Delta E}{kT}} \qquad (4)$$

Where, N is the number of excited atoms, N_0 is the number of atoms remaining in the ground state, A is a constant for a particular element, ΔE is the difference in energies of the two levels, k is the Boltzmann's constant and T is the temperature of flame.

Table 2. Flame temperature with different fuel and oxidant mixtures.

Fuel	Oxidant	Flame temperature (°C)
Propane	Air	1900
Propane	Oxygen	2800
H_2	Air	2100
H_2	Oxygen	2800
Acetylene	Air	2200
Acetylene	Oxygen	3000

The equation (4) gives that the fraction of atoms excited depends upon the temperature of the flame. Therefore, higher is the flame temperature, higher will be the number of excited atoms. The temperature of a flame is controlled by the type of fuel and the oxidant used. Some flame temperatures are given in Table 2.

2.2.3. *Components of Flame Photometer*

The main components of Flame Photometer are: (i) excitation unit (pressure regulator and flow meter for fuel gases, atomizer and burner), (ii) optical system (reflecting mirror, monochromator/filter), (iii) photosensitive detector and (iv) readout device (Fig. 5).

I. Flame or excitation unit: Depending upon the type of element to be analyzed, different burners and fuel-oxidant combinations are used to produce the analytical flame for burning the sample solution and exciting the atoms produced in the flame after burning. The liquid sample through a capillary tube is drawn into nebulizer, where it is converted into a fine mist or aerosol and passes into the flame which is produced by air (oxygen) and a fuel gas coming through side tubes. Then the sample in the form of mist is vaporized and converted into solid in the flame and subsequently, the atomization and excitation of the sample follows.

II. Optical system: Mirror (concave) is set behind the burner to reflect the radiations and to converge them at entrance of the monochromator/filter as the excited atoms emit radiations in all directions. Thus, the mirror helps in getting the maximum radiations to reach the detector. Monochromator or filter selects required wavelength from the continuous spectra emitted by the flame and allows them to pass through it. In a simple flame photometer, the monochromator is used which consist of an entrance slit, a dispersing material (a prism or grating) and an exit slit. The filter permits the light of the required wavelength but absorbs the light of other wavelengths wholly or partially. A monochromator also does the same function but instead of absorbing the undesired wavelengths, isolates the wide band of radiations with dispersing material and allows only the required wavelength to pass through the exit slit. A particular filter is used for specific analysis, therefore, different filters have to be used for analysis of different elements.

Filters are generally used when the emission spectrum of the elements contains a few spectral lines. In such cases, the wide wavelength ranges are allowed to enter the detector without affecting the sensitivity of the elemental analysis.

III. Detector: The radiations coming from the optical system is allowed to fall on the detector, which measures the intensity of radiation falling on it. The detector should be sensitive to radiations of all wavelengths that may be examined. In good flame photometers, the photomultiplier detectors are employed, which produce an electrical signal from the radiations falling on them.

IV. Readout device: The read out device gives output for the intensity of radiations emitted from the sample element. Presently most of the readout systems give digital out put.

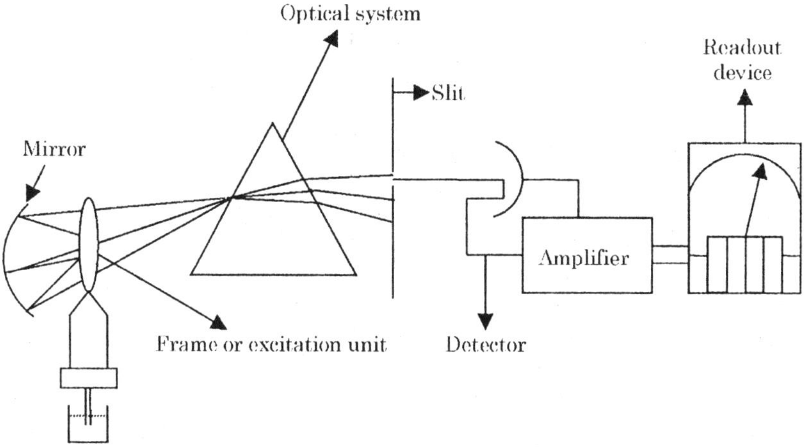

Fig. 5 : Diagram of flame photometer

2.2.4. *Operating instructions*

As modern flame photometers are generally operated with computers, therefore, one should carefully read the instruction manual of the instrument before starting the elemental analysis. The operational steps involved may vary with the type of instrument but the common steps are as follows:

(i) Switch ON the flame photometer.

(ii) Select the filter depending upon the element to be analyzed in the sample i.e. Na, K, Ca or Li. Open the lid of the chambers insert the required filter and close the lid.

(iii) Connect the air outlet nozzle of compressor and the fuel outlet nozzle of the fuel source to the air inlet and gas inlet nozzles, respectively, at the rear side of the instrument with pressure pipes.

(iv) Connect the capillary to the nebulizer inlet and ensure that the nebulizer is clean. Use cleaning wire, if needed. Connect the drain pipe to the overflow outlet of the collecting tank.

(v) Switch the compressor ON. Air is allowed into the mixing chamber. Adjust air control to read 0.75 kg cm^{-2} (10 psi) on the pressure gauge. This pressure may vary slightly from instrument to instrument.

(vi) Allow the fuel gas to enter the burner slowly and ignited to get flame close to the burner with 10 distinct sharp blue cones.

(vii) Allow the flame to stabilize for about 5 minutes.

(viii) Keep the sample solutions of required concentrations ready for measurement.

(ix) Dip the free end of the capillary in distilled water or the reagent blank and set the instrument readout display at zero.

(x) Now dip the free end of the capillary in the standard solution of highest concentration of the elements to be determined and adjust the output reading at 100.

(xi) Repeat the above two steps (9 and 10) to ensure that the zero and 100 are displayed when the blank and the standard solution of highest concentration, respectively, are aspirated into flame.

(xii) Insert the free end of the capillary in distilled water for a minute or two to wash the mixing chamber thoroughly before the actual sample measurement.

(xiii) Dip the free end of capillary in the test sample and read the concentration as displayed.

(xiv) Feed the standard solution of known concentration from time to time to check the reproduceability of the instrument, if large number of samples are to be analyzed in succession. Check the zero with the blank.

(xv) If the displayed reading of the sample exceeds that of standard solution of highest concentration used for calibration, dilute the sample suitably to have the concentration reading within the set range (0-100).

(xvi) At the completion of the measurements, feed distilled water to the nebulizer for 5 minutes for cleaning the system.

(xvii) To switch off the instrument, first cut off fuel supply to the burner and then switch off the compressor.

2.2.5. *Preparation of working curve*

Flame photometer gives the intensity of the emitted radiation as an output reading corresponding to the concentration of the element in the sample. To prepare the working curve (straight line) for an element, first the standard solutions are prepared and then the intensity values of the emitted radiations are plotted against these concentrations of the element (Fig. 6). The concentration of the element in the test sample is then obtained from the working curve corresponding to the intensity value of emitted radiation measured by the instrument for the test sample.

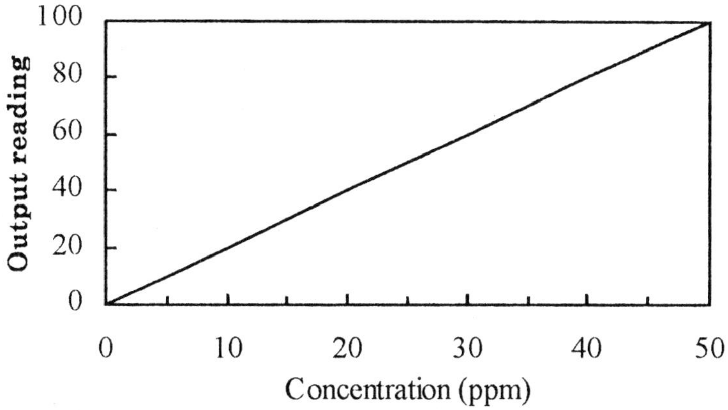

Fig. 6. Working curve for elemental analysis of flame photometer

2.3. Atomic Absorption Spectrophotometer

2.3.1. Introduction

It is the most powerful instrument to determine the elemental composition of the matter. It can determine 60-70 elements (including common rare earth metals) in concentrations ranging from trace to macro-quantities. The determination of micro-nutrients (Zn, Fe, Mn, B, Mo, Co, Cu, Ni) and heavy-metals (Cd, Cr, Pb) contents in plant tissue requires sufficiently higher temperature for their excitation, therefore, can not be estimated by flame photometer but require atomic absorption spectrophotometer.

2.3.2. Principle

When a solution containing metallic salt is introduced into a flame, the liquid is vapourized, leaving the particles of the solid salt. The salt is then dissociated into the gaseous state to give free neutral atoms. Some of the metal atoms may be raised to an energy level sufficiently high to emit the characteristic radiation of the metal — a phenomenon utilized in flame photometry. But a large fraction of the metal atoms (99%) will remain in unexcited state (non-emitting ground state). These unexcited atoms of a particular element are receptive of light radiation of their own characteristic wave length (the same wavelength as they would emit if excited). When a light of this wavelength is allowed to pass through a flame having atoms of the metallic salt, a part of that light will be absorbed and the absorption will be proportional to the density of the atoms in the flame. In atomic absorption spectrophotometer, we determine the amount of light absorbed or the absorption of energy by free atoms of test element in the gaseous state.

Mathematically, the total amount of light absorbed to calculate the concentration of the metallic element is expressed as:

$$\text{Amount of light absorbed} = \frac{\pi e^2}{mc} Nf$$

Where, e is the charge on the electron of mass m, c the velocity of light, N the total number of metal atoms capable of absorbing radiation of frequency υ in the light path and f the ability for each atom to absorb the radiation of frequency υ.

2.3.3. *Components of atomic absorption spectrophotometer*

An atomic absorption spectrophotometer consists of six major components, namely : radiation source, nebulizer, flame, monochromator, radiation detector and readout device (Fig. 7).

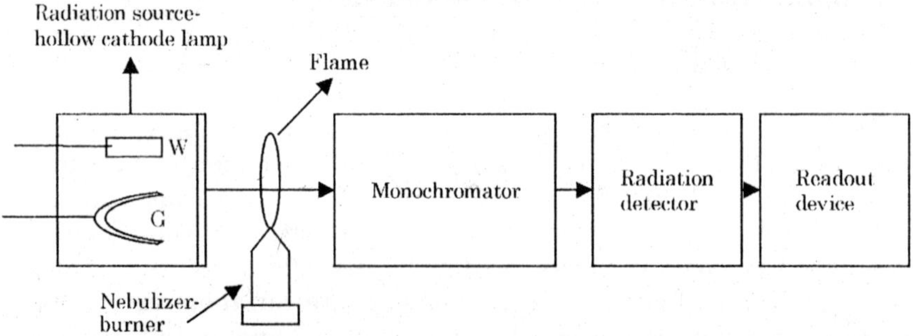

Fig. 7. Schematic diagram of an Atomic Absorption Spectrophotometer (AAS)

*I. **Radiation source:*** In atomic absorption spectrophotometer, a hollow cathode lamp is used as the source of radiation. In Fig. 7, A is anode and C is cathode terminated in hollow cup G, and both are sealed in glass tube filled with argon or neon gas at low pressure (hundreds of Pascal). W is a window of desired properties. The anode is generally a tungsten wire and the cathode is made up of an element whose spectrum is desired (test element When an electric potential (300-500V) is applied across the anode and cathode, electrons ionize neon or argon. Some of the ions possess enough energy to strip atoms away from the cathode and convert into an atomic gas on the surface of the cathode. Further collision of the gaseous/vaporized metal atoms from the cathode with the filled gas ions result in excitement of metal atoms which emit the characteristic spectrum of the metal used to construct the cathode. For example, using M(S) for the element M in its metallic (solid) state and M(G) when it is in its atomic state, emission corresponds to a series of steps:

$$M(S) \xrightarrow{Ne^+} M(G)^* \longrightarrow M(G) + \text{photon}$$

Thus, emission spectrum produced by the hollow cathode lamp is a sharp line spectrum of cathode material and the filled gas.

The neon or argon gas filled in the hollow cathode lamp performs three functions:

(i) Acts as main source of current-carrying capacity in the hollow cathode.

(ii) Dislodges atoms from the surface of the cathode

(iii) Causes excitation of the ground state metal atoms.

The spectral lines produced by the hollow cathode lamp are so narrow that they are completely absorbed by the atoms. By this method, one can easily detect and measure the atomic absorption.

Each hollow cathode lamp, emits the spectrum of that metal of which the cathode is made-up of. For example, copper cathode emits the copper spectrum; zinc cathode emits the zinc spectrum and so on. At the same time, the narrow spectral lines emitted by copper cathode are only absorbed by the copper atoms present in the sample to be analyzed by atomic absorption spectroscopy. For this reason, a different hollow cathode lamp has to be used for each element to be analyzed by atomic absorption spectroscopy which is not very convenient.

II. Nebulizer: Before the liquid sample enters the burner, it is converted into small droplets. The method of formation of small droplets from the liquid sample is called nebulisation. The gas emerging from the orifice of the burner with a high velocity causes such a high back pressure at the tip of the burner that aspirates the liquid sample through capillary and forms its fine droplets (mist).

III. Flame and burner: A flame is used for converting the liquid sample into gaseous state and also for conversion of the molecule entities into an atomic vapour. Two types of burners are commonly used : the total consumption burner and the premixed burner.

(i). Total consumption burner : Here the sample solution, the fuel and oxidizing gases are passed through separate passages to meet at the opening of the base of the flame. As the sample is a liquid, the flame breaks up the liquid sample into droplets which are then evaporated or burnt, leaving the residue which is reduced to atoms. The consumption burners use oxygen with hydrogen or acetylene and give very hot flames.

(ii) Pre-mixed burner : Mixture of the liquid sample and premixed (C_2H_2+O_2) is made to meet at the opening of the base of the flame. The premixed burner is suitable for the atomic absorption studies of metals of groups IA, IB and IIB, and others such as Pb, Mn and Ni, etc. Depending on the test element, either air- C_2H_2 or N_2O-C_2H_2 gas mixture is used. N_2O-C_2H_2 gives hotter flame as compared to air-C_2H_2 gas mixture.

IV. Monochromator: Monochromator selects the given absorbing line from spectral line emitted from the hollow cathode. The most common monochromators used in atomic absorption measurements are Prisms and Gratings.

V. Radiation detector: Detector measures the amount of absorption of radiation (photon). To analyze the wide range of elements, the detectors should cover wide spectrum range from Arsenic (As) line at 193.7 nm to Caesium (Cs) line at 852.1 nm. The spectral sensitivity of photomultipliers depends on the photo-sensitive material used for cathode of photocell. These materials are usually alloys of alkali metals with antimony, bismuth and /or silver.

VI. Readout device: Readout devices record the detected radiations. In modern instruments, the readout devices record data in digital form, which avoids the possibilities of errors in taking the reading on the meter or the recorder.

2.3.4. *Calibration curve for Atomic Absorption Spectrophotometer*

Calibration curve is a plot of the readings of absorbance recorded by Atomic Absorption Spectrophotometer (AAS) against the standard concentrations of the element to be determined in the test solution. Standard solutions of the element to be measured are prepared and the corresponding absorbance readings are recorded to draw a calibration curve. Calibration curve is then used to calculate the concentration of the test element. Pure compounds (AR grade) are used for preparing the standard solutions. Generally, a stock solution of higher concentration of the element to be determined, for example, 50, 100 or 1000 ppm is prepared carefully. Readymade preparations (pellets/tablets) of such compound in optimum sizes (quantities) be dissolved in suitable volumes for yielding standard stock solution of desired strengths, are also available in the market. The standard solutions of desired concentrations are then made by diluting the stock solution. The standard solutions should be stored in polyethylene or Teflon bottle. Glass should be avoided especially for low concentrations. Usually, at least three standards and a blank are used to cover the optimum working range, i.e., the concentration range of an element that will produce a range of 0.1-0.8 absorbance.

The calibration is performed by using the blank solution to set the instrument to read zero absorbance. The standards are then analyzed with the lowest concentration first and the blank should run between the standards to ensure that the baseline (zero point) remains unchanged. Samples are then analyzed one by one and their absorbance is recorded. A graph of absorbance versus concentration is plotted and a typical curve is shown in Fig. 8. Theoretically, linear relationship exists between the concentration and absorbance to a certain limit of concentration as per Beer's Law. The above calibration can also be performed in the concentration mode in which the concentration of the sample is readout directly.

Traditional unit to define the concentration of a sample in AAS (Atomic Absorption Spectrophotomter) was parts per million, (ppm). Currently accepted unit is $\mu g\ mL^{-1}$ or $mg\ L^{-1}$ although ppm is still widely used.

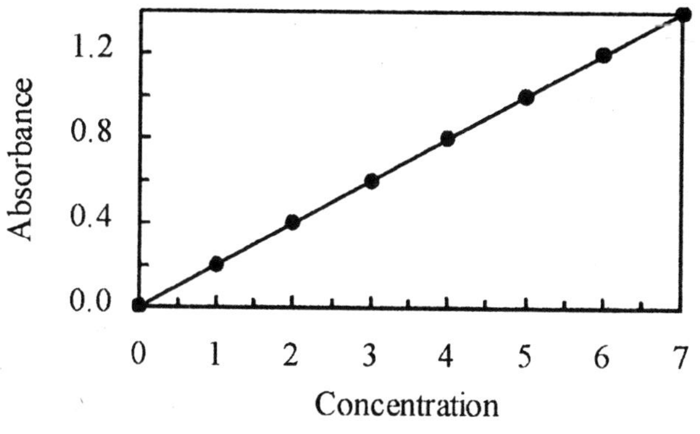

Fig. 8. The standard curve of an element for AAS showing conformity to Beer's law

2.3.5. *Operating instructions*

The modern atomic absorption spectrophotometers are operated with computers and the operating parameter vary with the model. Therefore, one should read the Instruction Manual of the instrument before starting the elemental analysis. The instrument must be optimized in terms of selection of the radiation source, fuel and oxidant flow rates for flame and burner adjustment etc., before carrying out analysis. The general steps are:

(i) Switch on the instrument.

(ii) Insert hollow cathode lamp of the element to be analyzed.

(iii) Adjust the required air and fuel mixture for the element to be analyzed.

(iv) Select the specific wavelength for the element to be analyzed.

(v) Calibrate the instrument with standard solutions.

(vi) Aspirate the test sample into the flame

(vii) The instrument will give concentration of the element in the sample directly.

(viii) After the analysis work is completed, give the final feeding with distilled water.

(ix) To switch off the instrument, first cutoff the fuel supply and then the air supply.

2.3.6. *Differences between flame photometer and atomic absorption spectrophotometer*

Table 3. Differences between flame photometer and atomic absorption spectrophotometer

S.N.	Flame photometer	Atomic abosrption spectrophotometer (AAS)
1	In flame photometer, the radiations emitted by the excited atoms are measured	In atomic absorption spectrophotometer, the characteristic radiations of the source (cathode lamp) absorbed by the unexcited atoms of liquid sample are measured.
2	The number of excited atoms in given concentration of the sample is relatively much less than the unexcited atoms. Hence, the degree of precision is low.	The number of unexcited atoms in a given concentration of sample, which absorb the energy of the source, is much higher than the excited ones. Hence, the degree of precision is much higher.
3	The radiations are emitted from the atoms excited by the thermal energy of the flame, therefore, it is greatly influenced by variation of flame temperature	The atoms absorb characteristics radiations produced from the cathode lamp and not from the flame. The flame is used only to produce the free atoms from the liquid sample. Hence, AAS is independent of flame temperature.
4	The spectral interference is probable in flame photometer	The spectral interference does not take place in AAS, because the atoms of a particular element can only absorb radiation of their own characteristic wavelength.

5	Most suitable for elements such as Ca, Mg, Na and K.	Most suitable for heavy metal elements of micro-nutrient (Zn, Fe, Cu, Mn, Co, Ni) and other toxic metals (Cr, Cd, Pb).

2.3.7. *Disadvantages of Atomic Absorption Spectrophotometer*

The disadvantage of the atomic adsorption spectrophotometer are (i) a separate cathode lamp is required for each element to be determined and (ii) estimation of elements like Al, Mo, Si, etc., which produce oxides in flame cannot be done.

2.3.8. *Probable Interferences in Atomic Absorption Spectrophotometer and Flame Photometer*

Various factors may affect the flame emission and atomic absorption measurements in elemental analysis.

I. Spectral interferences: In flame photometer, spectral interferences with the radiations emitted by the test element may arise from close proximity of wavelength of radiations emitted by other elements. These extraneous radiations, which may overlap partially or completely are read together with the requisite radiations in proportion to the degree of overlapping. This mainly occurs because, in practice, a chosen wavelength has in fact a finite band width. This difficulty, however, can be eliminated or minimized by increasing the resolution of the instrument, e.g., by using a prism rather than a filter. Spectral interference due to overlap of absorption radiations in AAS is rare because spectral width of an absorption radiation is about 0.005 nm.

II. Self absorption: Self absorption is a phenomenon of absorption (by the unexcited atoms) of the emitted radiation as they pass outwards from the central region of the flame. It occurs because central core of the flame is hotter than the surrounding outer cooler fringe of the flame and radiant energy can be absorbed through collision with atoms of its own kind present in the ground energy level. Such interference is particularly noticeable for the radiations originated from the lowest excited level and increases with the concentration of the sample solution. The self absorption interference can be minimized by diluting the sample solution.

III. Chemical interferences: A portion of the neutral atoms formed in the flame may combine to form new stable gaseous compounds and thereby decrease the concentration of neutral atoms and hence, their intensity is also decreased. For example, aluminates, silicates and phosphates are formed due to reactions between alkaline earth metals and oxy-anions and oxides of aluminium, vanadium, boron, etc. These interferences may be overcome by using a flame of high temperature enough to cause dissociation of the compound(s) to its (their) constituent atoms. For example, ions of phosphate, aluminate, etc., interfere with the determination of alkaline earth metals in the air/acetylene flame of low temperature, but this problem gets minimized at higher temperature nitrous oxide/ acetylene flame.

IV. Ionization interference: When the temperature of flame is increased, the intensity of emitted radiation is also increased. But this is not true in case of Li, Na or, which ionize easily. When these elements are heated to high temperature in flames (nitrous oxide/ acetylene), they become ionized instead of excited. Therefore, for such elements moderate temperature flames are required, which produce only excited atoms. The use of lower temperature flames may help in reducing ionization interferences. This interference can also be overcome by adding easily ionizable element to both standard and test sample.

V. Matrix interferences: The presence of large amounts of one or more elements in the test sample may significantly interfere with the determination of the test element. This interference effect can be overcome by adding equal amounts of interfering elements in the standard solutions as well, while preparing the standard curve.

3. REFERENCES

Chatwal, G. and Anand, S. (1995). *Spectroscopy-Atomic and Molecular* (Eds. M. Arora and S. Puri). Himalaya Publishing House, Bombay.

Isaac, R.A. and Kerber, J.D. (1971). Atomic absorption and flame photometry: Techniques and uses in soil, plant and water analysis. In: *Instrumental Methods for Analysis of Soils and Plant Tissue* (Ed. L.M. Walsh). pp 17-37. Soil Science Society of America, Madison, Wisconsin.

Willard, H.H., Merritt, L.L. Jr. and Dean, J.A. (1965). *Instrumental Methods of Analysis.* The D. Van Nostrand Company Inc., Princeton, New Jersey, USA.

Research Methods in Plant Sciences : Allelopathy.
Volume 4. Plant Analysis
Eds. S.S. Narwal, O.P. Sangwan and O.P. Dhankhar
Scientific Publishers (India), Jodhpur (2007) (pp. 39-48)

3

PLANT SAMPLES : COLLECTION, PROCESSING AND STORAGE

D.K. Bhandari, J.C. Sharma, S.S. Dahiya and M.L. Batra

Department of Soil Science, CCS Haryana Agricultural University, Hisar-125 004, India

1. INTRODUCTION

Proper collection of plant sample is as important as the reliable methods for their analysis. Unless a sample is truly representative, the analytical results are of little or no value. Likewise, the chemical analysis becomes meaningless, if there is an error in collection and preparation of plant samples. Hence, the collection and preparation of plant samples should be done with utmost perfection. Knowledge of nutrients concentration in growing plants can serve as a tool for correcting any deficiencies.

2. COLLECTION OF PLANT SAMPLES

2.1. Selection of Plant Parts

For a meaningful plant analysis, selection of right plant part, stage of growth and time of sampling are very important. It would be wrong and wasteful to collect any growing plant part, at any time for analysis. Plant sampling should be planned depending on the purpose of analysis. Plant parts (index tissue) recommended to be sampled for analyses have been identified (Table 1). The index tissue sampling is designed for nutrient diagnosis, monitoring and efficient nutrient management for optimum yield and excellent quality. However, the details of sampling in the field will depend upon the type of material and the purpose for which the sample is required. When more than one plant is sampled, include sufficient number of plants in each sample, to ensure that it adequately represents average composition of all plants sampled. The number of samples depends upon the variability in composition of plants.

Table 1. Plant parts recommended for sampling for nutrients anaysis.

Category	Crop	Part to be sampled with stage/age
Grains, Pulses, Oilseeds, Fibre and Commercial Crops	Wheat	Flag-leaf, before head emergence
	Rice	3rd leaf from apex, at tillering
	Maize	Ear-leaf, before tasseling
	Barley	Flag-leaf at head emergence
	Oat	Flag-leaf, before inflorescence emergence
	Pulses	Recently matured leaf, at bloom initiation
	Groundnut	Recently matured leaflets at maximum tillering
	Sunflower	Youngest mature leaf blade, at initiation of flowering
	Mustard	Recently matured leaf, at bloom initiation
	Soybean	3rd leaf from top, after 2 months of planting
	Cotton	Petiole, 4th leaf from the apex, at initiation of flowering
	Jute	Recently matured leaf, at 60 days age
	Sugarcane	3rd leaf from top after 3-5 months of planting
	Sugar beet	Petiole of youngest mature leaf at 50-80 days age
	Tea	3rd leaf from tip of young shoots
	Coffee	3rd or 4th pair of leaf from apex of lateral shoots, at bloom
Vegetables	Potato	Most recent, fully developed leaf (half-grown)
	Tomato	Leaves adjacent to inflorescence (mid-bloom)
	Onion	Top non-white portion (1/3 to 1/2 grown)
	Brinjal	Blade of most recent fully developed leaves

	Beans	Uppermost, fully developed leaves
	Cauliflower	Most recent, fully matured leaf, at heading
	Cabbage	Wrapper leaf at 2-3 months age
	Pea	Leaflets from most recent, fully developed leaves at first bloom
	Carrot	Most recent, fully matured leaf, at mid-growth
	Radish	Most recent, fully developed leaf
	Turnip	Most recent, fully developed leaf
	Beetroot	At 30-50 days age
	Spinach	5th leaf from tip (omit unfurled) at the stage of bud starting to small fruits
	Cucumber	Most recent, fully developed leaf
Ornamental plants	Bougainvillea*	Most recent, fully developed leaves at the stage of flower bud of pea size
Fruit crops	Almond	3rd leaf from top, at the beginning of bloom
	Apple and Pear	Leaves from middle of terminal shoot growth, 8-12th weeks after full bloom, 2-4 weeks after formation of terminal buds in bearing trees
	Blackberry	Latest matured leaf from non-tipped canes,4-6th weeks after peak bloom
	Cherry	Fully expanded leaves, mid shoot current growth in July-August
	Peach	Mid-shoot leaves, fruiting or non-fruiting spurs, mid-summer leaves, fruiting
	Strawberry	Youngest, fully expanded matured leaf without petiole, at peak or harvest period
	Plum	Leaves from middle of current season's extension growth, in January-February
	Banana	Petiole of 3rd open leaf from apex, 4 months after planting
	Cashew	4th leaf from tip of matured branches, at beginning of flowering
	Grapes	5th petiole from base at bud differentiation for yield and petiole opposite to bloom for quality
	Citrus	3-5th month old leaves from new flush, 1st leaf of the shoot, in June
	Guava	3rd pair of recently matured leaves, at bloom (August or December)

Mango	Leaf with petiole (4-7 month old) from middle of shoot
Papaya	6th petiole from apex, six months after planting
Pineapple	Middle 1/3rd portion of white basal portion of 4th leaf from apex, at 4-6 months stage
Pomegranate	8th leaf from apex at bud differentiation in April and August
Phalsa	4th leaf from apex, one month after pruning
Indian jujube	6th leaf from apex from secondary or tertiary shoot, 2 months after pruning

* Bougainvillea Jasmine, Croton, Fern, Ficus, Geranium, Gerbera, Gladiolus, Lilly, Orchid, Hibiscus and Rose

Sources: Kensworthy (1964), Tandon (1984), Reuter and Robinson (1986), Jones *et al.* (1991), Bhargav and Raghupati (1993), Singh *et al.* (1999) and Antil *et al.* (2000).

To study the influence of stage of growth, cultural practices, manurial treatment or grazing on the mineral composition of a crop or pasture, the whole plant material from selected area will be required. Whereas, to compare the composition of different plant species growing on various soil types, only pure species samples should be taken at comparative stage of growth. With respect to grazing experiments, cut the pasture samples, 1.0-1.5 cm above the ground level, to simulate the grazing animal. Cutting of herbage closer to ground level, destroys the plant and contaminates the sample with soil. On each plot, many areas (3-5) are selected randomly and the herbage cut from each is lightly packed into separate open mesh bags and transferred to the laboratory as quickly as possible. The herbage of each sample is thoroughly mixed and divided into a number of sub-samples such that it weighs approximately 100 g (fresh weight). The sample is first air dried, then dried at 60°C in an hot air oven.

To collect a composite sample for whole shoot analysis from a field, it is usually taken by cutting the shoots (aerial part) with stainless scissors from small representative areas and then combining the whole material as one sample. To determine the effect of soil type on the plant composition, such samples are restricted to a small area typical of the field and a composite soil sample is taken to represent the same area. In case of plant sample collection from various parts, separate the different plant parts immediately, wash the sample with tap water to remove dust particles. Put the sample in an oven and dry at 60°C for 36 hours. If the roots are to be studied, then before taking root sample, adequately irrigate the plant site and then uproot the whole plant carefully from the wet soil, retaining even the fine/active roots. Gently dip the plant roots in water several times to remove the adhering soil, as far as possible.

2.2. Precautions during Sample Collection

(i) **Avoid soil contamination** : The contamination of plant sample with soil etc. must be avoided. For example, a small amount of contamination with a calcareous soil would cause significant error in the values for calcium and magnesium in the plant sample, while the reverse is true for iron and aluminium. To determine manganese, copper, zinc and boron, strict precautions must be taken to avoid contamination with soil.

(ii) Plant should be cut individually 1-3 cm above the ground level. However, height at which the plant sample can be cut depends on the material being harvested and the purpose of sampling. Likewise, tall crops, (cereals etc.) can be cut at a somewhat greater height than pasture samples.

(iii) Avoid taking samples after top dressing of the fertilizers or pesticides. Because their significant amounts may be retained on the leaves of the plants as a fine dust. Amount of trace elements normally present in plants are so small that such error may be considerable. In cereals and grasses, fine dust from the applied fertilizer, may collect between the leaf sheath and the stem and persist for long time. To avoid all possibility of contamination, plant samples should not be transported in proximity to fertilizers or pesticides containing the elements under investigation.

3. PREPARATION AND PROCESSING OF PLANT SAMPLES

3.1. Preparation of Samples

After collection of plant samples, fresh tissues should be made free from dust and other foreign material by washing either with distilled water or double distilled water if micronutrients analysis is to be done. Place the washed samples on filter paper sheets for soaking the moisture followed by putting them in separate open mesh (perforated bags) for air drying for 1-3 days. Then put the plant samples in new paper bags to dry in a forced air oven at 65°C ± 2°C for 48 h. Grind the samples in an electric stainless steel grinder using 0.5 mm sieve. Put each sample in oven and dry again for few hours more for constant weight and then store them in paper bags for further analysis.

3.2. Processing of Samples

Processing of samples includes their (i) drying and (ii) grinding.

3.2.1. *Drying of samples*

After collection, plant samples should be dried as rapidly as possible to reduce chemical and biological changes to minimum. Considerable loss in dry weight may occur due to respiration, while proteins are also broken down to simpler nitrogenous compounds, if drying is unduly delayed. When fresh material has to be transported long distances to the laboratory, it should be lightly packed in loosely woven bags prior to transportation. Tight packing may lead to considerable overheating, so increasing respiration losses & other changes.

I. Precautions during sample drying

- (i) Green material should be dried in a well ventilated oven so that the water vapours are rapidly removed from the immediate vicinity of the sample. The material to be dried, should be very loosely packed in open shallow trays, otherwise free movements of the moisture laden air is impeded. Poor ventilation leads to considerable decomposition of some organic constituents of the sample. Rapidly dried samples are much greener in colour than those badly dried.
- (ii) Drying is most conveniently done at 60-70^{0}C temperature.
- (iii) Do not air dry plant samples in the proximity of chemicals or chemical solutions.
- (iv) Samples for determination of Cu or Zn should not be dried in drying ovens made of copper or brass, unless these ovens are completely protected with a suitable synthetic varnish coat to prevent metallic contamination of sample.
- (v) Drying ovens made of stainless steel are particularly suitable for the drying of samples for most trace elements determination (Jackson, 1967).

II. Determination of moisture

Since the analysis results are expressed on oven dry weight basis, thus it is necessary to determine the moisture content of air dried tissue (Wilde *et al.,* 1979). Duplicate determination should be made on each sample of plant tissue. Results of air dried tissue analysis are then converted to the oven dry basis. One or 2 g sample of ground air dried tissue is dried in an oven at 105°C for 12 to 16 h. The samples are cooled in desiccators and weighed. The difference in weight represents loss of moisture and is expressed as a percentage of oven dry weight.

3.2.2. *Grinding of plant samples*

Before grinding, ensure that the plant sample is completely dry and bulky samples are reduced by quartering until a convenient amount of material is obtained. The material (plant) is carefully separated into leaf, stem, seed, etc., when large samples are sufficiently reduced by quartering; the sub-sample is chaffed (cut in small pieces) and further reduced by quartering until an amount of 150-250 g is obtained. The material is then ready for grinding. It gives a more homogeneous sample.

The fineness of grinding depends upon the size of sub-sample to be taken for chemical analysis. For many purposes it is sufficient to grind the material until it passes through a stainless steel sieve with round holes of 1 mm dia. If amounts of only 1-2 g are needed for chemical determination, it should be ground to pass 0.5 mm sieve.

3.2.3. *Grinding mill*

Two types of mills are recommended for grinding plant samples : (i) Wiley mill and (ii) Christy and Norris Junior (C&N mill). Grinding is best done on the oven dry material but except in very humid localities, both of these mills will grind air dry samples very satisfactorily (Piper, 1966).

(i) Wiley Mill: In the Wiley mill, grinding is done by the action of a set of steel knives revolving against another set of knives mounted in the circular outer casing of the grinding chamber. One of the three inter-changeable brass sieves (2 mm, 1 mm and 0.5 mm round holes) can be fitted into the lower part of this outer casing. Material to be ground is fed at the top of the mill and carried around between the blades of the revolving and stationary knives until it is sufficiently fine to drop through the holes of the sieve at the bottom. The finely ground material is collected below the sieve. This mill grinds plant material with minimum heating.

(ii) C & N Junior Mill : The grinding is done by a four armed beater cross which revolves freely at high speed (12,500 r.p.m.) in a plane circular grinding chamber. At this high speed any plant material, fed through the feed inlet is shattered or torn until it is sufficiently fine to pass through a small iron screen plate clamped in the periphery of the grinding chamber. Several interchangeable screen plates are supplied with the mill. Those with holes 0.4 mm and 1 mm in diameter are most useful. Owing to the centrifugal action of this mill, a considerable amount of air is drawn through it when in operation. It is, therefore, necessary to use a 'dust tight receptacle' to collect the ground sample. It is valuable, when only small amount of material are available for grinding.

3.2.4. *Precautions during grinding*

Following precautions must be observed during the grinding of plant materials :

(i) In steel mill, do not grind samples in which iron is to be determined, to prevent contamination. Grind such material in C&N mill made of bronze, phosphor bronze or other non-ferrous alloy.

(ii) Steel mills are permissible for manganese and cobalt determination

(iii) When copper or zinc is to be determined, avoid contamination from brass (it contains copper and zinc).

(iv) Wide stemmed funnels used for transferring ground samples to the sample bottles should be made of either glass or aluminium.

(v) When copper and iron are to be determined, two separate samples must be ground in the appropriate mills to avoid their contamination.

(vi) Do not grind the samples in vicinity of chemical/chemical solutions used in the laboratory.

4. STORAGE OF PROCESSED PLANT SAMPLES

Most plant samples are collected for immediate analysis. After grinding, the contents of plant samples are mixed thoroughly, transferred to the suitable bottles or polythene bags, labeled clearly and packed properly for storage in a moisture free environment. Care must be taken that samples are stored away from the chemicals or chemical solutions and fertilizer(s).

5. REFERENCES

Antil, R.S., Singh, A. and Dahiya, S.S. (2002). *Practical Manual for Soil and Plants Analysis*, Department of Soils, CCS Haryana Agricultural University, Hisar.

Bhargava, B.S. and Raghupati, H.B. (1993). Analysis of plant materials for macro and micronutrients. In : *Methods of Analysis of Soils, Plants,*

Waters and Fertilizers (Ed., H.L.S. Tandon), pp 49-82. Fertilizer Development and Consultation Organization, New Delhi.

Jackson, M.L. (1967). *Soil Chemical Analysis*. Asia Publishing House, Bombay.

Jones, J.B. Jr., Wolf, B. and Mills, H.A. (1991). *Plant Analysis Handbook: A Practical Sampling Preparation, Analysis, and Interpretation Guide*. Micro-macro Publishing Inc. Athens, Georgia.

Kensworthy, A.L. (1964). *Fruit, Nut and Plantation Crops – A Guide For Collection of Foliar Samples for Nutrient Element Analysis*. Mimeographed, Horticulture Department, Michigan State University, East Lansing, USA. pp. 39.

Piper, C. S. (1966). *Soil and Plant Analysis*. Hans Publishers, Bombay.

Reuter, D.J. and Robinson, J.B. (1986). *Plant Analysis – An Interpretation Manual*. Inkata Press, Melbourne, Sydney.

Singh, D., Chhonkar, P.K. and Pandey, R.N. (1999). *Soil, Plant and Water Analysis: A Methods Manual*. Indian Agricultural Research Institute, New Delhi, pp viii + 160.

Tandon, H.L.S. (1984). *Methods of Analysis of Soil, Plant, Water and Fertilizer*. Fertilizer Development and Consultation Organization, New Delhi.

Wildey, S.A., Corey, R.B., Iyers, J.G. and Voigt, G.K. (1979). *Soil and Plant Analysis for Tree Culture*. Oxford & IBH, New Delhi.

Research Methods in Plant Sciences : Allelopathy.
Volume 4. Plant Analysis
Eds. S.S. Narwal, O.P. Sangwan and O.P. Dhankhar
Scientific Publishers (India), Jodhpur (2007) (pp. 49-62)

4

DIGESTION OF PLANT SAMPLES

O.P. Sangwan, Deepak Khatak and D.S. Dahiya

Department of Soil Science, CCS Haryana Agricultural University, Hisar - 125 004, India

1. INTRODUCTION

Digestion of plant samples is the pre-requisite step to determine total nutrients present in plants. The precision and interpretation of plant analysis, results depends on how accurately the plant samples are digested, after their proper collection and processing (cleaning, drying and grinding) (Kensworthy, 1964; Tandon, 1984; Jones *et al* 1991). In plants, the nutrients exist in various organic combinations such as carbohydrates, proteins, fats, etc. and digestion of plant material releases them in mineral forms. The weight of plant sample for digestion varies from 0.2 to 2.0 g. depending on (i) the type of plant species, (ii) stage of plant growth and (iii) the plant parts. However, the usual weight of plant sample taken for digestion for total elemental analysis is 0.5 g. Digestion of plant samples can be done by (a) dry ashing or (b) wet oxidation. Dry ashing is done by igniting the plant material in a muffle furnace at 550-600 °C followed by the extraction of ignited material in dilute acid (HCl or HNO_3). Whereas, wet oxidation involves digestion of plant material in a mixture of two or three concentrated oxidizing acids ($HClO_4$, H_2SO_4

and HNO_3) at a specified high temperature. The details of techniques and procedures for dry and wet oxidation of plant samples adopted for soil and plant analysis are available (Jackson, 1973; Chopra and Kanwar, 1991; Greweling, 1976; Tandon, 1993; Chapman and Prat, 1961; Piper, 1966; Gupta, 1999). We have described main methods of plant digestion for analysis of macro-, micronutrients and toxic elements.

2. METHODS OF DIGESTION

There are two methods for digestion of plant samples (i) dry ashing and (ii) wet oxidation. Each of these methods have merits and demerits. The choice of method depends on the availability of equipments, reagents and number of samples to be analyzed. Earlier, dry ashing was used commonly but presently it is restricted to the laboratories which are not so well equipped. Wet digestion is more common method in well equipped modern laboratories, with provision of exhausting the acid fumes (fume hoods) produced during the digestion.

2.1. Dry Ashing of Plant Samples

The dry ashing technique is simple and free from the problem of acid fumes. It is used for the analysis of K, Na, Ca, Mg, Cu, Zn, Fe, Mn, Mo and B (Issac and Jhonson, 1975). It is further sub-divided as (a) simple dry ashing and (b) dry ashing with basic compound such as magnesium nitrate. The nutrient elements such as Ca, Mg, K, Na, and most of the trace elements, which do not volatilize at high ignition temperature (around 550 °C) can be digested by simple dry ashing. Whereas others like N, P and S, which volatilize at high ignition temperature should not be digested by dry ashing and they should be wet digested. However for S, dry ashing can be done with magnesium nitrate, to avoid the volatilization losses (Greweling, 1976).

Advantages

It is simple, free from problem of fumes of boiling acids and is preferred method for B and Mo estimation.

Disadvantages

It is slow and can not be used for N, P and S estimation because they get volatilized at high ignition temperature (550°C) required for ashing.

2.1.1. *Simple Dry Ashing Method (For elements except S and P)*

Principle

The plant samples for the determination of nutrient elements (K, Na, Ca, Mg, Cu, Zn, Fe, Mn, Mo and B), which do not volatilize at high temperature can be digested by dry ashing in furnace using suitable silica, porcelain or platinum crucible and extracting them in dilute HNO_3.

Apparatus

Muffle furnace, silica/porcelain/platinum crucible of 50mL capacity, electric hot plate, 50mL volumetric flask, 1L volumetric flask, 1L beaker, wash bottle, etc.

Reagents

(i) **Concentrated Hydrochloric Acid (HCl)**: Transfer 100 mL AR grade concentrated HCl in 250 mL beaker.

(ii) **5M Nitric acid (HNO_3):** Transfer 310 mL of concentrated HNO_3 (16N) in one litre volumetric flask and make the volume with distilled water and store in reagent bottle.

(iii) **2M Nitric acid (HNO_3):** Transfer 125 mL concen-trated HNO_3 (16N) in one litre volumetric flask and make the volume with distilled water and store in reagent bottle.

Note: To prepare standard solutions of different concentrations (Normality/Molarity) see Chapter-1 of this book.

Procedure

(i) Take 0.5 g well dried and ground plant sample in a crucible and place the crucible inside the muffle furnace.

(ii) Raise the temperature of muffle furnace slowly and then fix at 550 °C for 4 to 6 hours, so that the entire plant material turns to white ash i.e., no black ash is left.

(iii) The crucible is then removed from the furnace and cooled. The contents of crucible are then dissolved in 5 mL of 2M HNO_3 and transferred to 50 mL volumetric flask.

(iv) Makeup the volume with distilled water and allow it to stand for some time for the silica to settle down. Then filter and store this aliquot for subsequent elemental analysis.

(v) If large amount of visible black ash remains (after step-ii) in the crucible, to remove it add 2-3 mL of 5M HNO_3 to cooled ash and evaporate the sample to dryness on an electric hot plate set at 100-120 °C. Then place it in the muffle furnace and heat it for about 1hour at 500 °C.

(vi) Remove the crucible from the muffle furnace and after cooling moisten the ash with few drops of distilled water and then add 2 mL of concentrated HCl. Evaporate it to dryness on a hot plate (100-120 °C) and continue to heat it for another1/2 to 1 hour. Then follow the rest of the procedure as explained under step-iii above to finally obtain 50mL aliquot required for analysis.

2.1.2. *Dry Ashing of Plant Samples with Magnesium Nitrate (For Sulphur)*

Elements like sulphur, phosphorus and chlorine are volatile elements (at high ignition temperature of 550 °C) hence, cannot be determined correctly by dry ashing as done for K, Na, Ca, Mg, Cu, Zn, Fe, Mn, B and Mo. To prevent the volatilization losses of S during dry ashing process, some basic compounds such as magnesium nitrate, sodium carbonate and sodium peroxide are used with the plant samples. The method of dry ashing for S estimation is described below.

Principle

Dry ashing of plant samples for S estimation can be done by using magnesium nitrate ($Mg(NO_3)_2$), which prevents the losses of S by volatilization at high ignition temperature (550 °C) required for ashing in the muffle furnace. The organic form of plant S is converted into sulphate form on ashing, which is extracted in HCl acidified warm water.

Apparatus

Muffle furnace, porcelain crucible of 50 mL capacity, electric hot plate, 50 mL volumetric flask, 1.0 volumetric flask, 250 mL and 1.0 beakers, wash bottle, etc.

Reagents

(i) **Magnesium Nitrate Solution:** Weigh and transfer 955 g of $Mg(NO_3)_2.6H_2O$ to 1 Litre beaker and then add 350 mL of distilled water and one g of MgO in it. Boil the solution and filter through Whatman filter paper No. 42 & make the final volume to one litre with distilled water.

(ii) **Concentrated Hydrochloric Acid (HCl):** Transfer 100 mL AR grade concentrated HCl in 250 mL beaker.

Procedure

(i) Take 0.5 g of finely ground plant sample in a 50 mL porcelain crucible and add 5 ml of $Mg(NO_3)_2$ solution to it. Now swirl the contents gently. Heat the crucible on hot plate at about 180 °C till dryness of the contents. Then transfer the hot crucible to the furnace.

(ii) Slowly raise the temperature of the furnace to 500 °C and ignite the contents of the crucible until entire carbon gets oxidized. Remove the crucible from the furnace and allow to cool.

(iii) Add 5 ml concentrated HCl to the crucible, evaporate to dryness on hot plate at 110-120 °C and bake it for one hour to dehydrate silica.

(iv) Add few drops of distilled water to moisten the ash and then add one ml of concentrated HCl and warm it on hot plate to dissolve the salts.

(v) Transfer the content of the crucible to a 50 mL volumetric flask and make up the volume with warm distilled water. Allow the silica to settle down. If the solution is not clear, add few drops of concentrated HCl till the solution becomes clear. This solution contains total S as well as other elements of plant sample oxidized to mineral form.

2.1.3. *Dry Ashing with Sodium Peroxide (For Sulphur)*

The plant sample is ignited in a Nickel crucible at high temperature (400-500 °C) with an oxidative mixture of Na_2CO_3 and Na_2O_2 in a muffle furnace resulting in the oxidation of organic form of S to sulphate form, which is extracted with acidified (HCl) distilled water.

Apparatus

100 mL Nickel crucible, Ni or Pt rod, electric hot plate, 600 mL beaker, 100 mL volumetric flask, aliquot storage bottles, wash bottle, etc.

Reagents

(i) **Sodium carbonate (Na_2CO_3)**: This is used as anhydrous powder during the digestion. For its convenient use, transfer approximately 20-30 g. of this salt in a 50 mL beaker.

(ii) **Sodium peroxide (Na_2O_2)**: This is also used as anhydrous powder during the digestion.

(iii) **Concentrated Hydrochloric acid (HCl)**: Transfer 100 mL AR grade concentrated HCl in 250 mL beaker.

Procedure

(i) Weigh 0.5 g of finely ground plant sample in a 100 mL Nikel crucible.

(ii) Add 2.5 g anhydrous Na_2CO_3 to the sample and mix thoroughly with the Ni or Pt rod and moisten with 1 ml distilled water.

(iii) To the above mixture, add Na_2O_2 taken in splits of 0.25 g at a time. Mix the material thoroughly after addition of each split and continue addition until mixture becomes nearly dry and granular.

(iv) Place the crucible on electric hot plate and heat carefully, stirring it occasionally, until contents are fused. Care should be taken that the material does not get ignited at this stage, otherwise experiment will fail.

(v) Remove the crucible from hot plate and allow it to cool. Then cover the hardened mass in the crucible with more Na_2O_2 up to depth of about 5 mm.

(vi) Heat crucible first gradually and then raise the temperature of the hot plate until fusion takes place again. Rotate crucible occasionally to bring the total contents in contact with oxidizing material.

(vii) Continue heating for 10 minutes after fusion is complete.

(viii) Cool for some time and then place the warm crucible along with contents into a 600 mL beaker and add about 60 mL distilled water carefully.

(ix) When initial violent action ceases, wash the material out of crucible. Make it slightly acidic with HCl by adding small portions at a time.

(x) Transfer the material into 100 mL volumetric flask. Cool the contents of flask and make the volume. Filter the contents with Whatman No. 42 filter paper.

(xi) The filtered solution contains total S as well as other elements of the plant oxidized to mineral form.

2.1.4. *Precautions for Dry Ashing*

(i) The plant samples to be dry ashed, must be fully dry and weighed accurately.

(ii) Avoid use of silica crucible (which may cause sticking of nutrient elements like Cu on its walls). To avoid this problem, use porcelain or platinum crucible to avoid this defect.

(iii) The temperature of the furnace should be increased slowly to reach its ignition mark of 550 °C.

(iv) The temperature of the muffle furnace should not be allowed to go beyond 550 °C ± 15 °C.

2.2. Wet Digestion of Plant Samples

Wet digestion of plant samples employs the oxidation of plant material by using suitable combination of two or three concentrated oxidizing acids ($HClO_4$, HNO_3 and H2SO_4 taken in specific ratios) (Table 1). The type of acid mixtures to be used for digestion will depend on the type of nutrient elements to be determined in the plant samples. The suitability of oxidizing acid mixtures for different nutrients is given in Table 1.

Table 1. Suitability of oxidising acid mixtures for different nutrients.

S. N.	Acid Mixture	Elements to be determined
1	HNO_3+$HClO_4$	All nutrients*except N
2.	H_2SO_4+H_2O_2	Only for N
3.	H_2SO_4+HClO4	All nutrients* except S
4.	HNO_3 + H_2SO_4+$HClO_4$	All nutrients* except N and S.

*All nutrients do not include Cl and Mo in the present case.

Principle

The basic principle of wet digestion is that a known weight of plant material is treated with the specified mixture of two or three concentrated oxidizing acids at boiling temperatures resulting in release of the plant nutrient elements from their original organically bound forms to the mineral forms.

2.2.1. *Wet Digestion by Diacid Mixture of HNO_3 and $HClO_4$ (All nutrients, except N)*

This method is most suitable for the analysis of P, K, S, Ca, Mg, Na, Al, Se and all micronutrients except B and Mo in plant material. A known weight of plant sample is oxidized by the di-acid mixture of HNO_3 and $HClO_4$ (4 : 1 Ratio). The digest is diluted with distilled water to obtain an optimum volume of aliquot.

Advantages : It is easier, less time consuming and convenient to analyse large number of samples.

Disadvantages: The boiling acids produce fumes, which is problem in ordinary laboratories.

Apparatus

Analytical balance, electric hot plate, 50 mL conical flask, acid fume hood, tongs, 5 mL tilt measure, 500 mL beakers, dropper, 50 mL volumetric flask, funnels, whatman No.1 filter paper, wash bottle, aliquot storage bottles, etc.

Reagents

(i). Concentrated Nitric acid (HNO_3, 16N)

(ii). Concentrated Perchloric acid ($HClO_4$, 11.6N)

Di-acid mixture : Prepare the di-acid mixture by mixing the concentrated HNO_3 and $HClO_4$ in the ratio of 4:1, respectively. To prepare 1 litre of the di-acid mixture at a time, mix 800 mL of concentrated HNO_3 in 200 mL of concentrated $HClO_4$ taken in a 2 litre beaker. Allow the mixture to cool and store it in a reagent bottle.

Procedure

(i) Weigh 0.5 g well dried and powdered plant material in a 50 mL conical flask.

(ii) Add 5 mL diacid mixture and allow it to stand over night or about 2-3 hours for pre-digestion.

(iii) Place the flask on the electric hot plate and start heating for digestion at 180 ^{o}C -200 ^{o}C until dense white fumes of acids are evolved and a clear colorless solution/ aliquot is obtained. If the solution is not clear, remove the flask from hot plate cool for some time and add 1-2 mL of the diacid mixture and heat it again to obtain a clear solution.

(iv) Remove the flask from the hot plate when a little amount (0.5 mL) of the digest is still left. The contents of the flask should not be allowed to totally dry out.

(v) Allow the contents of the flask to cool down and then add approximately 20 mL distilled water to dissolve the contents.

(vi) Transfer the entire contents of the digestion flask to a 50 mLvolumetric flask and make up the volume with distilled water.

(vii) Filter the aliquot through Whatman No. 1 filter paper and store in a plastic aliquot storage bottle for its subsequent analysis.

2.2.2. *Wet Digestion with H_2SO_4 and H_2O_2 (Only for Total N)*

In this method the plant sample is charred with concentrated H_2SO_4 and then complete oxidation is accomplished by use of H_2O_2. To estimate the total nitrogen in plants, ground plant sample of known weight can be digested with concentrated H_2SO_4 taken in a Kjeldahl flask with excess of 30% H_2O_2 and K_2SO_4-Catalyst mixture, which, give nascent oxygen and raise the boiling point of the digestion system.

Apparatus

Analytical balance, Kjeldahl digestion-set, 250 mL Kjeldahl flask, 10 mL-tilt measure, 50 mL volumetric flask, 500 mL beaker, plastic aliquot storage bottles, wash bottle, etc.

Reagents

(i) Concentrated Sulphuric acid (H_2SO_4)-36N

(ii) Hydrogen Peroxide (H_2O_2)-30%

(iii) **Potassium Sulphate-Catalyst Mixture:** This mixture is prepared by mixing 10g potassium sulphate (K_2SO_4) and 1 g catalyst mixture (20 parts of $CuSO_4$ and 1 part of Se powder).

Procedure

(i) Take 0.5 g dry and grinded plant sample and transfer in 250 mL Kjeldahl flask.

(ii) Add 20 mL concentrated H_2SO_4 slowly and carefully with the help of tilt measure such that whole of the plant material is well covered at the bottom. Allow it to stand over night or at least 2-3 hours for pre-digestion.

(iii) Place the flask in the Kjeldahl digestion-set and heat the sample at 150-200 °C for about 15-20 minutes. The contents of the flask would turn from black to brown.

(iv) Remove the Kjeldahl flask from the digestion set and cool for 10-15 minutes. Then add 1-1.5 mL of H_2O_2 and place the flask in the digestion set for another 5-10 minutes. Repeat this process 2-3 times or unless the solution becomes clear.

(v) Add 1g of potassium sulphate-catalyst mixture as listed above in reagents and heat the contents of flask to complete the digestion at high temperature (360 °C).

(vi) After completion of digestion, cool and transfer the entire contents of flask to a 50 mL volumetric flask and make up the volume with distilled water.

(vii) Filter the aliquot through Whatman No. 1 filter paper and store in a plastic aliquot storage bottle for its subsequent analysis.

2.2.3. *Wet Digestion by Modified Kjeldahl Method (For Total-N including NO_3- N)*

In simple Kjeldahl method, only a fraction of plant NO_3 will be reduced to ammonia during digestion. The mature plants have low contents of NO_3. However, if plants samples are taken at an early stage of growth or have received high rates of applied N, there may be substantial quantities of NO_3 present in plant tissue, which would not get reduced to ammonia during digestion. So if the NO_3 is present in substantial quantities, the digestion must be modified (Bremner and Mulvany, 1982) as explained under.

Principle

Nitro compounds formed by the reaction of salicylic acid with NO_3 in acid medium are reduced to corresponding amino compounds by heating the mixture with sodium thiosulphate and zinc dust. The main product of nitration is 5-nitrosalicylic acid and small amount of 3-nitrosalicylic acid. Drying of plant samples in oven eliminates the interference of water in analysis as nitration of salicylic acid does not take place if more than a trace of water is present.

Apparatus

Analytical balance, Kjeldahl digestion-set, 250 mL Kjeldahl flask, 10 mL-tilt measure, 50 mL volumetric flask, 500mL beaker, plastic aliquot storage bottles, wash bottle, etc.

Reagents

(i) **Sulphuric salicylic acid:** Dissolve 1g of salicylic acid in 30 mL of concentrated H_2SO_4.

(ii) **Sodium thiosulphate:** 24 mesh dried powder.

(iii) **Potassium Sulphate-Catalyst Mixture:** This mixture is prepared by mixing 10g potassium sulphate (K_2SO_4) and 1 g catalyst mixture which contains 20 parts of $CuSO_4$ and 1 part of Se powder.

Procedure

(i) Weigh 0.5 g dried plant tissue and transfer it to a 100 mL Kjeldahl flask.

(ii) Add 20 mL Sulphuric salicylic acid mixture in the flask and swirled so as to let the dry samples come in contact with the reagents.

(iii) Allow it to stand overnight.

(iv) Add 5 g of sodium thiosulphate powder and heat gently for about 5 minutes avoiding frothing. Now cool the contents and add 10g of potassium sulphate-catalyst mixture. Again put on Kjeldahl apparatus for digestion at full heat (360 °C-400 °C) until the solution is clear. The digestion is continued for 1 hour after the solution is cleared.

(v) After completion of digestion, digest is cooled and diluted with distilled water to make the volume of 50 mL.

2.2.4. *Wet Digestion by Diacid Mixture of H_2SO_4 and $HClO_4$ (All nutrients except S)*

A known weight of plant sample taken in a conical flask is treated with excess of di-acid mixture of H_2SO_4:$HClO_4$ (9:1) and digested on electric hot plate at 180°C-200°C. The digest is diluted with distilled water to obtain an optimum volume of aliquot.

Apparatus

Analytical balance, electric hot plate, 50 mL conical flask, acid fume hood, tongs, 5 mL tilt measure, 500 mL beakers, dropper, 50mL volumetric flask, funnels, Whatman No. 1 filter paper, wash bottle, aliquot storage bottles, etc.

Reagents

(i). Concentrated Sulphuric acid (H_2SO_4, 36 N)

(ii). Concentrated Perchloric acid ($HClO_4$, 11.6N)

Prepare the di-acid mixture by mixing the concentrated H_2SO_4 and $HClO_4$ in the ratio of 9:1, respectively. To prepare 1 litre of the di-acid mixture at a time, mix 900 mL of concentrated H_2SO_4 in 100 mL of concentrated $HClO_4$taken in a 2 litre beaker. Allow the mixture to cool and store in a reagent bottle.

Procedure

(i) Weigh 0.5 g dry and powdered plant material in a 50 ml conical flask.

(ii) Add 5 ml di-acid mixture and allow it to stand over night or for at least 3-4 hours for pre-digestion.

(iii) Place the flask on the electric hot plate and start heating for digestion at 180 °C -200 °C until dense white fumes of acids are evolved and a clear solution/ aliquot is obtained. If the solution is not clear remove the flask from hot plate cool for some time and then add 1-2 mL of the diacid mixture and heat it again to obtain the clear solution.

(iv) Remove the flask from the hot plate when a little amount (0.5mL) of the digest is still left. The contents of the flask should not be allowed to totally dry out.

(v) Allow the contents of the flask to cool down and then add approximately 20 mL of distilled water to dissolve the contents.

(vi) Transfer the entire contents of the digestion flask to a 50 mLvolumetric flask and make up the volume with distilled water.

(vii) Filter the aliquot through Whatman No. 1 filter paper and store it in plastic aliquot storage bottle for its subsequent analysis.

2.2.5. *Wet Digestion by Triacid Mixture of HNO_3, H_2SO_4 and $HClO_4$ (All nutrients except N and S)*

The digestion of plant material with a tri-acid mixture of concentrated Nitric acid (HNO_3), Sulphuric acid (H_2SO_4) and Perchloric acid ($HClO_4$) is specially recommended for the estimation of P and K. A known weight of plant sample can be digested with excess of tri-acid mixture of HNO_3:H_2SO_4:$HClO_4$ (10:1:4) taken in a digestion flask on an electric hot plate at a temperature of 180°C-200°C. The digest is diluted with distilled water to obtain an optimum volume of aliquot.

Apparatus

Conical flasks (100 ml capacity), electric hot plate, volumetric flasks, funnels, beakers, measuring cylinder, aliquot storage bottles, Whatman No.1 filter paper, etc.

Reagents

(i). Concentrated Nitric acid (HNO_3, 16N)

(ii). Concentrated Sulphuric acid (H_2SO_4, 36N)

(iii). Concentrated Perchloric acid ($HClO_4$, 11.6N)

Prepare the tri-acid mixture by mixing HNO_3, H_2SO_4 and $HClO_4$ acids, in ratio of 10:1:4 (v/v), respectively, in a beaker. To prepare 1.5 litre of the tri-acid mixture at a time we can take and mix 1 litre of HNO_3, 100 ml of H_2SO_4 and 400 ml of $HClO_4$ in a 2 litre beaker. Allow the mixture to cool and then store in a glass reagent bottle.

Procedure

(i) Weight 0.5 g dry and powdered plant material in a 100 ml conical flask.

(ii) Add 10 ml of the tri-acid mixture and allow it to stand over night or about 2-3 hours for predigesting.

(iii) Place the flask on the electric hot plate and start heating for digestion at 180 °C-200 °C until dense white fumes of acids are evolved and a clear colorless solution/ aliquot is obtained.

(iv) Remove the flask from the hot plate when about 1-2 ml of the digest is still left. The contents of the flask should not be allowed to totally dry out.

(v) Allow the contents of the flask to cool down and then add approximately 20 ml of distilled water to dissolve the contents.

(vi) Transfer the entire contents of the digestion flask to a 50 ml volumetric flask and make up the volume with distilled water.

(vii) Filter the aliquot through Whatman No. 1 filter paper and store it in a plastic bottle for its subsequent analysis of P and K.

2.2.6. *Precautions*

(i) Always use well dried and powdered plant samples.

(ii) Select a proper diacid or triacid mixture for digestion of plant material depending on the nutrient element(s) to be estimated in the plant samples.

(iii) Use suitable devices like tilt measure or droppers for adding the concentrated acids or acid mixtures in the digestion flask.

(iv) Avoid the contact of acid fumes to any part of the body. Use proper kits.

(v) To avoid effect of fumes always use digestion chambers provided with fume hoods.

(vi) The digestion flask should be removed from the hot plate, when 1-2 mL of the digest is still left at the bottom to avoid charring.

3. REFERENCES

Bremner, J.M. and Mulvaney, C.S. (1982). Nitrogen-total. In: *Methods of Soil Analysis-Part 2. Chemical and Microbiological Properties* (Ed., A.L. Page *et al.*). *Agronomy Monograph* No. **9**: 595-624. American Society of Agronomy, Madison, WI, USA.

Chapman, H.D. and Prat, P.F. (1961). *Methods of Analysis of Soils, Plants and Water*. University of California, USA.

Chopra, S.L. and Kanwar, J.S. (1976). *Analytical Agricultural Chemistry*. Kalyani Publishers, New Delhi.

Greweling, T. (1976). *Chemical Analysis of Plant Tissues.* Agronomy No.6. Cornel University Agricultural Experimental Station, Ithaca, NY. *Search Agriculture* **6(8)**:1-35.

Gupta, P.K. (1999). *Soil, Plant, Water and Fertilizer Analysis*. Agro. Botanica, Bikaner.

Issac, R.A. and Johnson, W.C. (1985). Elemental analysis of plant tissue by plasma emission spectroscopy: collaborative study. *Journal Association of Analytical Chemists* **58**:436-440.

Jackson, M.L.(1973). *Soil Chemical Analysis*. Prentice-Hall of India, New Delhi.

Jones Jr., J.B., Wolf, B. and Mills, H.A. (1991). *Plant Analysis Handbook: A Practical Sampling, Preparation, Analysis and Interpretation Guide.* Micro-Macro Publishing Inc. Athens, Georgia, USA.

Kensworthy, A.L. 1964. Fruit, Nut and Plantation Crops – A Guide for Collection of Foliar Samples for Nutrient Element Analysis. Department of Horticulture, Michigan State Univ. East Lansing USA. pp.39.

Piper, C. S. (1966). *Soil and Plant Analysis*. Hans Publishers, Bombay.

Tandon, H.L.S. (1993). *Methods of Analysis of Soils, Plants, Water and Fertilizers* (ed.). Fertilizer Development and Consultation Organization, New Delhi.

Section II.

DETERMINATION PLANT NUTRIENTS

Research Methods in Plant Sciences : Allelopathy.
Volume 4. Plant Analysis
Eds. S.S. Narwal, O.P. Sangwan and O.P. Dhankhar
Scientific Publishers (India), Jodhpur (2007) (pp. 65-68)

5

CARBON

O.P. Sangwan and V.K. Phogat

Department of Soil Science, CCS Haryana Agricultural University,
Hisar-125 004, India

1. INTRODUCTION

Plants absorb 15 essential mineral nutrients (N, P, K, Ca, Mg, S, Zn, Fe, Cu, Mn, Co, Ni, B, Mo and Cl) from soil and C and O directly from air. They derive H either directly or indirectly from the soil water. About 94-99.5% of fresh plant tissues are made up of C, H and O and only 0.5 to 6% is constituted by all mineral nutrients. Despite their small proportion, the mineral nutrients usually limits the growth and development of crop plants. Plant growth (except in drought, extreme cold, poor drainage or disease) is not critically retarded by lack of C, H and O. It is the mineral nutrients derived from the soil, which greatly affects the plant growth, during their deficiency. That is why the entire emphasis is on mineral nutrition and hardly on C, H and O, although the later constitutes bulk of plant body.

2. COMPOSITION OF PLANTS

In green plant material, about 75% is water and rest 25% is dry matter (Waksman, 1948). Carbon, hydrogen and oxygen (i.e. C,H,O), dominate the bulk of tissues of plant. On weight basis, the dry matter is mainly composed of carbon (44%), oxygen (40%) and hydrogen (8%) amounting to 92%. The mineral matter (ash contents) comprising of all essential nutrient elements is about 8% only (Table 1).

Table 1. Plant materials composition on weight basis.

Constituent compounds		Constituent compounds	
Name	Content (%)	Name	Content (%)
1. Carbohydrates	60	1. Organic fractions	92
Sugar and starches	1-5	Carbon	44
Hemicelluloses	10-30	Oxygen	40
Celluloses	20-50	Hydrogen	08
2. Proteins	10	2. Mineral fractions	08
Crude Proteins	2-15	(N,P, K, Ca, Mg, S, Fe,	
True proteins	1-10	Cu, Mn, Zn, Co, B, Mo,	
		Cl, Ni)	
3. Facts, oils, waxes etc.	5		
4. Lignins	25		

Source : Waksman, 1948

Among the organic compounds, carbohydrates (sugars, starches, hemicelluloses and celluloses) constitute 60%. Lignin is more complex than carbohydrates and is found in older woody plant tissues. Fats, oils and waxes are found in seeds especially of oil seed crops. The others constituent compounds of plant system are proteins mainly made of C, H, O and N and smaller amounts of S, Mn, Cu, Fe, etc.

Plant absorb essential nutrient elements in some specific ionic forms (e.g. NO^{-}_{3}, NH^{+}_{4} are the available forms of N in soil) and are assimilated and metabolized in the plant system and present in the plants as carbohydrates, proteins, etc.

3. DETERMINATION OF TOTAL CARBON

The plant sample to be analysed should be dry as the results of the analysis are expressed on oven dry weight basis (Wildey *et al.*, 1979).

Dry combustion method

The dry combustion method (Nelson and Sommer, 1982) to estimate total carbon in the plant tissues is simple and quick method, hence, commonly recommended.

Principle

Total carbon in a plant sample can be determined by oxidizing it by dry combustion in a furnace at medium temperature range (525°C-550°C). The loss in weight of oven dry plant material due to combustion is measured as total C.

Apparatus

Platinum crucible, muffle furnace, electric balance, oven, and desiccator.

Procedure

1. Take a clean and well dried platinum crucible and note down its weight on electric balance.
2. Transfer in it 5 g oven dry plant material.
3. Place the crucible in the muffle furnace and slowly increase the temperature to 550°C, till the white ash is obtained.
4. Remove the crucible from furnace, and cool it in a desiccator and weight it.
5. Note the loss of weight in plant material due to combustion.

Calculations

Let the weight of the platinum crucible be = W_1 g

The weight of the oven dry plant material = 5 g

The weight of the crucible + ignited material devoid of C = W_2 g

Thus total weight of C (loss of weight on combustion) = 5- (W_2-W_1) g

Hence, % total carbon in plant (oven dry basis) = $\frac{5-(W_2-W_1)}{5}100$

4. REFERENCES

Nelson, D.W. and Sommer, L.E. (1982) A rapid and accurate procedure for estimation of organic carbon in soil. *Proceeding Indian Academy of Science* **84**: 456-462.

Waksman, S.A. (1948). Composition of plant residues. In : *Nature and Properties of Soils* (Ed. Nyle C. Brady) Prentice Hall of India, New Delhi, p. 280.

Wildey, S.A.; Corey, R.B.; Iyers, T.G. and Voigt, G.K. (1979). *Soil and Plant Analysis for Tree Culture*. Oxford & IBH Publishing Co., New Delhi.

Research Methods in Plant Sciences : Allelopathy. Volume 4. Plant Analysis
Eds. S.S. Narwal, O.P. Sangwan and O.P. Dhankhar
Scientific Publishers (India), Jodhpur (2007) (pp. 69-86)

6

NITROGEN

D.J. Dahiya, Ravi Kumar and M.S. Kuhad

Department of Forestry, CCS Haryana Agricultural University, Hisar-125004, India

1. INTRODUCTION

Nitrogen is an essential major plant nutrient and is absorbed by plants in large amounts. It is constituent of chlorophyll, protoplasm, protein and nucleic acids. Most crop plants absorb nitrogen as nitrate from soil, however, some plants also absorb in ammonium form. Application of nitrogen improves the quality of leafy vegetables, fodder and food-grains by increasing their protein content. In India, nitrogen deficiency has become wide spread due to intensive cropping. Nitrogen deficiency in plants decreases protein content and flowering in plants. Several methods of nitrogen determination in plant extract are available depending on sample condition, accuracy/ reproducibility of results and the facilities available.

2. METHODS TO DETERMINE NITROGEN

In traditional method of measuring nitrogen (Dumas, 1831), sample is ignited with copper oxide (CuO) at high temperature (600°C) in steam of purified carbon dioxide and the gaseous products are led over hot Cu to reduce nitrogen oxide to N_2 and then over CuO to convert CO into CO_2. The N-CO_2 mixture is collected in a nitrometer containing concentrated alkali, which absorb CO_2 and volume of N_2 is measured manometrically. This method did not remain popular owing to cumbersome apparatus and tedious experimental procedure.

Kjeldahl (1883) modified earlier method, involving conversion of organic nitrogen into ammonia by boiling with sulphuric acid. It is common method to determine nitrogen in plant sample and involves two steps:

(i). Digestion of sample to convert organic nitrogen to NH_4^+-N, and (ii). Determination of NH_4^+-N in the digest.

The digestion is done by boiling plant sample with concentrated di-acid mixture of sulphuric acid and perchloric acid that converts the organic-N into ammonical form. (The digestion procedure is described in Chapter 4 of this book). The nitrogen of the nitrogenous substances is converted into ammonia by boiling with concentrated sulphuric acid and the ammonia so formed reacts with sulphuric acid and forms ammonium sulphate as indicated by following reaction:

Organic-nitrogenous compound —

$$\text{Digestion } (360\text{-}410\ ^0\text{C}) \rightarrow NH_3$$

$$2NH_3 + H_2SO_4 \rightarrow (NH_4)_2SO_4$$

Once the digestion of plant sample is completed, the ammonium nitrogen produced is determined by any of four methods. (i) The first method depends upon alkaline distillation, the liberated ammonia is absorbed in acid and titrated. (ii) The second method depends upon the reaction of ammonia with color developing reagent to give a colored solution; (iii) third is micro-diffusion method and the (iv) fourth is by NH_4-ion electrode.

Table 1. Methods to determine NH_4^+-N in digested plant extract

Method name		NH_4-N range (mg/L)	Interference	Comments
Steam-distillation	i) with NaOH	0.1-1000	Volatile amide	Rapid, accurate, widely used
	ii) with MgO	0.1-1000		
Micro-diffusion with MgO		0.1-100	None	Simple
Colorimetric	i) Indophenol blue	0.005-20	Ca, Mg Ca, Mg,	Simple, rapid. i). Needs less amount of aliquot.
	ii) Nessler	0.1-10	Fe, SO_3	i). Distil sample before colour development
Ion-electrode		0.1-1400	Volatile amide	Sensitive, rapid

2.1. Alkaline steam distillation method

This methods based upon alkaline steam distillation of digested solution with a base (NaOH or MgO) and thus librated ammonia is absorbed in known volume of standard acid/ or in excess boric acid solution, which is measured by titrating against standard acid or base. This is common method being simple, rapid, and accurate. Various organic and inorganic substances that often interfere with colorimetric method do not affect it. This method is also applicable to colored solution.

Principle

Ammonium-N present in digested plant sample can be determined by distillation of known volume of digested aliquot of plant tissue with strong alkali. When nitrogen-containing solution is distilled with alkali (NaOH) solution, the ammonia gas expelled is trapped in excess of known volume of standard acid/ boric acid solution. The amount of ammonia absorbed is determined by titrating it with standard alkali/ or with standard acid, using methyl red-bromo-cresol green as mixed indicator.

Reaction

$$(NH_4)_2SO_4 + 2\,NaOH^- \rightarrow 2\,NH_3 + Na_2SO_4 + 2\,H_2O$$

$$NH_3 + HCl \rightarrow NH_4Cl$$

$$\text{Or } 2NH_3 + H_2SO_4 \rightarrow (NH_4)_2SO_4$$

Apparatus

1. Micro-Kjeldahl digesation unit
2. Steam distillation apparatus
3. Distillation flask of 50 mL
4. Microburette of 5 mL, graduated with 0.01 mL interval

Reagents

(i). **Sodium hydroxide (45%):** Dissolve 450 g of NaOH in one litre of distilled water.

(ii). **Boric acid –mixed indicator solution**: Dissolve 40 g of H_3BO_3 in about 900 mL of hot distilled water. Cool and add 20 mL of indicator solution (prepared by dissolving 0.5 g of bromocresol green and 0.1 g of methyl-red in 100 mL of 95% ethanol). Add 0.1 M of NaOH solution drop wise, until the colour of solution becomes reddish-purple. Make complete one litre volume by adding distilled water.

(iii). Standard H_2SO_4 or HCl (0.01 N)

(iv). Standard 0.01 N Na_2CO_3 or $BaCl_2$.

(v). **Methyl-red indicator**: Dissolve 0.2 g of methyl red in 50 mL of alcohol, add 50 mL of distilled water and filter the solution.

(vi). **Standard NH_4-N (50 ppm) solution**: To prepare 50 ppm of NH_4-N solution, dissolve 0.236 g of ammonium sulphate [$(NH_4)_2SO_4$] in 1000 mL of ammonia-free distilled water. Mix the solution thoroughly and store in refrigerator.

Procedure

Transfer 20 mL of digestion solution in 500 mL vacuum jacketed distillation flask after thoroughly swirling, attach the distillation flask to the distillation head. To absorb the evolved ammonia of the digestion solution, place a conical flask of 125 mL capacity containing 10 mL of 0.01 N H_2SO_4/ or boric acid solution below the discharging tube of condenser in such a way that the end of the ammonia delivery tube should be dipped into solution of the receiver flask. The distillation flask is blocked with rubber cork having double hole. Through one hole, the steam of condenser is passed having its end above the distillation mixture and through other hole, pass a glass tube dipping its end into the distillation solution and its other end being attached to a 5 cm long tube with a pinch-cork inserted into a funnel which is filled with excess of 40% NaOH (about 50 mL). After making all connections airtight, start distillation initially at low flame. Pour excess of 45% NaOH into the distillation flask through the rubber tube by opening the stopcock and immediately close the mouth of rubber tube with a pinch-cork to minimize the loss of ammonia through this passage. If anyhow double hole-cork is not available then pour the NaOH solution from a graduated cylinder into the distillation flask along the neck and connect the flask quickly to the distillation apparatus. In this way, the total sodium hydroxide is concentrated in a layer beneath the solution and there is no danger of escape of ammonia. Now gradually increase the flame upto boiling. When the distillate becomes ammonia-free (at about 40-50 mL distillate), disconnect the receiver flask and wash the end of delivery tube with distilled water and then put-off the burner to prevent back sucking. Also run a blank analysis by performing distillation of blank sample and absorbing liberated ammonia in 10 mL of 0.01 N H_2SO_4/Boric acid dilute standard acid.

The ammonium may be determined either by volumetric or a colorimetric method. In volumetric method, the titration of the absorbed ammonia can be done in two ways:

(a) Acid-base titration

The absorbed ammonia in known quantity (10 mL) of dilute standard H_2SO_4 or HCl (0.01 N) during distillation is being back titrated with standard NaOH or Na_2CO_3 (0.01 N) using methyl-red as indicator. At the end point, the color change is from lavender to colourless and then to green. Simultaneously, titration of blank sample is also carried out.

Calculations

1 mL of 1 N.NaOH or Na_2CO_3 = 0.014 g nitrogen

N (%) in plant sample = (B-T) x 0.014 x N x D x 100 / W x A

Where,

B= mL of standard NaOH/ Na_2CO_3 used for blank titration

T= mL of standard NaOH/ Na_2CO_3 used for plant sample

N= Normality of standard NaOH/ Na_2CO_3

D = Dilution factor of plant material

W = Weight of plant sample (g)

A = mL of aliquot taken for distillation

(b) Acid-acid titration

In this method, the liberated ammonia is absorbed in boric acid mixed indicator solution and is being back titrated against standard (0.01 N) HCl or H_2SO_4. At the end point, the blue colour of the solution disappears and solution turns pink with next drop of titrant. A blank titration is also performed simultaneously by taking distillate of blank solution.

$$NH_3 + H_3BO_3 \rightarrow NH_4^+ + H_2BO_3^-$$

$$H_2BO_3^- + H^+ \rightarrow H_3BO_3$$

To know the amount of NH_4-N derived from the reagents used for analysis, a control should be performed in each series of analysis by taking 10 mL of digested sample of blank (without plant material) in the distillation flask The blank determination is subtracted from each determination. To ensure the reliability of the methods, a distillation of standard solution and its titration for series of analyses should also be done with same reagents and procedure used for analyses of plant samples.

Calculations

1 mL of 1 N HCl or H_2SO_4 = 0.014 g nitrogen

N (%) in plant sample = (T-B) x 0.014 x N x D x 100 / W x A

Where, T= mL of standard acid used for plant sample titration

B = mL of standard acid used for blank sample titration

D = Dilution factor of plant material

N = Normality (0.01) of standard acid used

W = Weight of plant sample (g)

A = mL of aliquot taken for distillation

Advantages of acid-acid titration over acid-base titration

(i). Only one standard acid is required for titration, whereas in acid-base titration both the reagents (receiver and titrant) need standardization.

(ii). A large excess of boric acid can be used to ensure complete absorption of NH_3, even if the amount of NH_3 liberated is more than anticipated.

(iii). Use of boric acid with mixed indicator shows the presence of sufficient alkali in the digest both for the occurance and the completion of the NH_3 distillation.

2.2. Semi-micro Kjeldahl method (Bremner, 1965)

Principle

The determination of nitrogen in plant tissue consists of two stages. In the first stage, the organic nitrogen is converted into ammonium ions through the oxidation of organic nitrogen in the plant by boiling with concentrated sulphuric acid. The second stage consists of the determination of the amount of ammonium ions formed from the organic nitrogen.

Apparatus

1. Kjeldahl digestion rack with fume aspirator
2. Analytical balance
3. Micro-burette (10 mL)
4. 125-mL conical flask
5. Distillation set

Reagents

(i) Digestion mixture: Mix three parts of copper sulphate and one part of mixture of anhydrous sodium sulphate plus selenium taken in ratio of 5:1.

(ii) Concentrated sulphuric acid

(iii) Sodium hydroxide (45%): Dissolve 450 g of NaOH in one litre of distilled water.

(iv) Boric acid –mixed indicator solution: Dissolve 40 g of H_3BO_3 in about 900 mL of hot distilled water. Cool and add 20 mL of indicator solution (prepared by dissolving 0.5 g of bromocresol green and 0.1 g of methyl-red in 100 mL of 95% ethanol). Add 0.1 M of NaOH solution drop wise until the color of solution becomes reddish-purple. Make complete one litre volume by adding distilled water.

(v) Standard H_2SO_4 or HCl (0.01 N)

Procedure

Weigh 0.2-0.5 g (depending on the nitrogen content) of ground plant sample and transfer in a 100-mL Kjeldahl flask. Add about 3 g of digestion mixture to the Kjeldahl flask by long-stem funnel. From the dispensing burette, add 7 mL of concentrated H_2SO_4 by giving washing to neck of flask. Place the flask in the digestion unit and turn on the heat and suction pump. Start the digestion initially at low heat to prevent frothing. When the sample starts carbonizing, then swirl the flask gently at 5 min interval. After 15-20 min, the heat is gradually increased until the contents become clear. Do not allow flame to touch the flask above the part occupied by liquid; to prevent loss of NH_3 due to decomposition of $(NH_4)_2SO_4$. The digestion is continued for 20 min till the solution become colourless and white precipitate appears on cooling. At the end of digestion, the heating is stopped. After the flask is cooled, add 25 mL of NH_3-free distilled water carefully along the side of the flask and swirl the flask.

Transfer the complete digested solution to distillation flask by giving several rinsing with distilled water. Attach the distillation flask to the distillation head of entrance tube having three-way stopcock. The upper container of the entrance tube is filled with NaOH solution. Place a 125-mL Erlenmeyer flask containing 10 mL of boric acid - mixed indicator-solution at the end of delivery tube of condenser in such a way that the tip of delivery tube should dip into solution. Start the water to boil in the steam generator and allow the cold water to flow through the condenser. Pour the NaOH solution to distillation flask by opening the stopcock. Turn the three-way stopcock to permit the entrance of steam into the distillation flask. Allow the distillation to proceed until about 40-45 mL of distillate have collected in the receiver flask and the coming distillate becomes NH_3-free (test with red litmus). Disconnect the receiver flask by rinsing the condenser tube with distilled water. The three-way stopcock is turned off, to check the entry of steam and to allow the entry of air into the distillation flask, meanwhile, open the valve of the steam generator to prevent high pressure.

The ammonia absorbed in boric acid solution is back titrated with standard 0.01N H_2SO_4 or HCl contained in a 5-mL burette. At the end point, the blue colour just disappears and on next drop the solution turns to pink.

Calculations

1 mL of 1 N HCl or H_2SO_4 = 0.014 g nitrogen

N (%) in plant tissue = (T-B) x N x 0.014 x 100 / S

T= Sample titration, mL standard acid

B = Blank titration, mL standard acid

N = Normality of standard acid

S = sample weight (g)

Note: % crude protein in plant tissue=% N in plant tissue x 6.25

2.3. Magneous- oxide method (Bremner & Keeney, 1965)

Principle

The NH_4-N in plant sample is determined from the liberated ammonia by steam distillation of known volume of digested aliquot with finely powdered magneous-oxide (MgO). The liberated ammonia is absorbed either in standard acid solution or excess boric acid-indicator solution (H_2SO_4/HCl) that is estimated with titrating against either standard alkali solution or standard acid solution.

Apparatus: Steam distillation rack, pipettes, 125-mL conical flasks.

Reagents

(i). Magnesium oxide (light): Heat the MgO in an electic muffle furnace at 600-700 ^{0}C for 2.0 hours. Cool the product in desiccator having KOH pellet.

(ii). Boric acid –indicator solution: Prepare as described in section 2.1 of this chapter.

(iii). Standard ammonium (NH_4) solution: Prepare as described in section 2.1 of this chapter.

Procedure

Pipette 10 mL of aliquot of the digested plant sample into a distillation flask and dilute it to about 30-40 mL with distilled water. Take 5 mL of boric acid-indicator solution in to a 50-mL conical flask and place this flask under the condenser of steam distillation apparatus in such a way that the tube of delivery tube of condenser should dip into the solution. Now add about 0.2 g of MgO into distillation flask, through a dry powder funnel having a long stem that reaches down into the bulb of flask. Attach the distillation flask to the steam distillation apparatus immediately and commence steam distillation. When the distillate becomes about 30-35 mL in the receiver flask and become ammonia free (red litmus paper do not turn blue) then stop the distillation. First remove the receiver flask after rinsing the end of delivery tube with distilled water. Determine the NH_4-N in the distillate by titration with standard (0.01 N) H_2SO_4 contained in the burette. The colour change at the end point is from green to a permanent faint pink. The liberated ammonia can also be absorbed in standard (0.01 N) H_2SO_4 which can be estimated by back titration with standard (0.01 N) Na_2SO_4 using methyl-red as indicator. The liberated ammonia from blank solution (without plant sample) is absorbed in standard acid solution (or excess boric acid-indicator solution) by running a distillation of blank solution.

Calculations

1 mL of 1 N HCl or H_2SO_4 = 0.014 g nitrogen

N (%) in plant sample = (T-B) x 0.014 x N x D x 100 / W x A

Where, T= mL of standard acid used for plant sample titration

B = mL of standard acid used for blank sample titration

N = Normality (0.01) of standard acid used

D = Dilution factor of plant material

W = Weight of plant sample (g)

A = mL of aliquot taken for distillation

2.4. Micro-diffusion method (Bremner and Shaw, 1955)

Micro-diffusion method for determination of inorganic form of nitrogen offers the advantages for low cost and simplicity. The micro-diffusion unit consists of two chambers: an inner chamber for NH_3 absorbing solution (H_3BO_3) and an outer chamber for the soil sample extract. In diffusion method, the distillation and aeration are eliminated. This method is based on the principle that if a volatile substance is liberated in one vessel where it exerts a tension, it can be absorbed by simple gaseous diffusion into another chamber where its tension is reduced to zero or on the surface of the absorbing fluid.

Principle

In this method, the NH_3 liberated by the treatment of the digested plant tissue solution with potassium carbonate (placed in outer chamber) is absorbed in boric acid-indicator solution (placed in the inner chamber) and is determined by titration of this solution with standard H_2SO_4 (0.01N).

Apparatus

- Micro-diffusion dish of two chambers.
- Micro-burette of 5-mL capacity

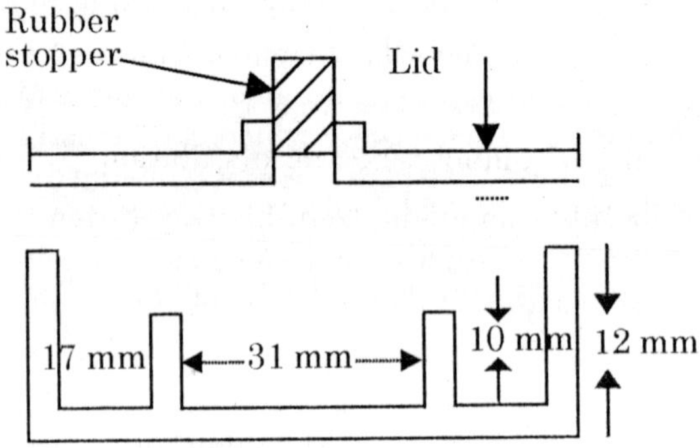

Figure 1. Micro-diffusion unit

Reagents

(i) Potassium carbonate (45%) solution: Dissolve 450 g of K_2CO_3 in one liter of distilled water. Add to it 4-5 drops of non-ionic wetting agent, Tergitol-NPX.

(ii) Boric acid-indicator solution: Prepare as described in section 2.1 of this chapter

(iii) Standard H_2SO_4 or HCl solution (0.01N).

(iv) Standard NH_4-N solution: Prepared as described in section 2.1 of this chapter

Procedure

Take 5 mL aliquot of the plant digest in the outer peripheral sample chamber and place 5 mL of boric acid-indicator solution into the central chamber. Add about 5 mL of K_2CO_3 solution in the aliquot solution. Seal the lid air-tightly and place the diffusion chamber into constant temperature chamber at 25 ^{0}C for 2 days (period required for complete diffusion of the ammonia into H_3BO_3 solution). Remove the lid and add 2-3 mL of distilled water to the central chamber containing boric acid-indicator solution. Determine the absorbed NH_4-N into boric acid solution by titration with standard H_2SO_4 (0.01N). The colour change at the end point is from green to a permanent, faint pink.

Calculation

1 mL of 1 N HCl or H_2SO_4 = 0.014 g nitrogen

N (%) in plant sample = (T-B) x 0.014 x N x D x 100/ W x A

Where, T= mL of standard acid used for plant sample titration

B = mL of standard acid used for blank sample titration

N = Normality (0.01) of standard acid used

D = Dilution factor of plant material

W = Weight of plant sample (g)

A = mL of aliquot taken for distillation

Disadvantages: (i) Relatively small volume of extract that can be safely analysed, limits the accuracy and precision especially with solutions containing low concentration of inorganic-N and (ii) Relatively long time period (2-3 days) required to obtain results.

2.5. Colorimetric method

Colorimetric methods are commonly used to analyse inorganic nitrogen in plant and soil extracts due to their sensitivity, speed and ease of use. Ammonium in plant extract is estimated in two step procedure involving (i) distillation and (ii) colour development by Nessler's or indophenol blue color method. These methods are based on the reaction of ammonia with the reagent to give a coloured solution. In this method, the procedure is same up to digestion as described earlier. The ammonia in digested solution can also be measured by spectrophotometer or colorimeter methods. Sometimes the presence of Fe, Al, Ca and Mg ions interfere with the colorimetric determination of ammonium by introducing turbidity in the solution, hence, it becomes essential to distill off ammonia from the digest plant solution. In most cases, sulphuric acid digestion of plant material contains interfering ions in very small amount, even then accurate determination of ammonium by colorimetric method is possible

Standard nitrogen solutions for curve calibration and checking analytical procedure:

(i) Standard NH4-N solution (50 ppm): Dissolve 0.236 g $(NH_4)_2SO_4$ in one liter of ammonia free distilled water and store in a refrigerator.

(ii) Standard (ammonium + nitrate)-nitrogen solution (50 ppm NH_4+-N and 50 ppm NO_3-N): Dissolve 0.236 g $(NH_4)_2SO_4$ and 0.361 g KNO_3 in one litre of ammonia and nitrate free distilled water and store in a refrigerator.

(iii) Standard (ammonium + nitrate + nitrite)-nitrogen solution (50 ppm NH_4+-N, 50 ppm NO_3 N and 25 ppm NO_2-N): Dissolve 0.236 g $(NH_4)_2SO_4$, 0.361 g KNO_3 and 0.123 g $NaNO_2$ in one litre of ammonia and nitrate free distilled water and store in a refrigerator.

2.5.1. *Indophenol blue method* (Kempers, 1974)

Principle

In alkaline pH solution, when ammonia reacts with phenol/ salicylate solution in the presence sodium hypochlorite, forms an intense blue color complex. The intensity of the color is proportional to the amount of ammonia involved, whose optical density can be measured by a colorimeter at 625 nm.

Apparatus: Spectrophotometer

Reagents

(i) **Phenol-nitroprusside reagents**: Dissolve 7 g of phenol and 34 mg of sodium nitroprusside [disodium pentacyanonitrosylferrate, $Na_2Fe(CN)_5NO.2H_20$] in 100 mL of NH_4-free distilled water. Mix the solution well and store in a dark colored bottle in a refrigerator.

(ii) **Buffered hydrochlorite reagent**: Dissolve 1.480 g of sodium hydroxide (NaOH) in about 70 mL of NH_4- free distilled water. Add to this 4.98 g of sodium mono-hydrogen phosphate (Na_2HPO_4) and 20 mL of hypochlorite (NaOCI) solution (5-5.25% NaOCI). Adjust the pH in between 11.4 to 12.2 by adding NaOH, if required to raise the pH. Dilute the solution to final 100 mL volume with distilled water.

(iii) **Ethylene diamine-tetra acetic** acid (EDTA) solution: Dissolve 6 g of ethylene-diamine-tetra acetic acid disodium salt (EDTA sodium) in 100 mL of distilled water. Mix the solution well and adjust the pH to 7.0.

(iv) **Standard NH_4- N (50ppm) solution**: Prepare as described in section 2.1 of this chapter.

Procedure

Pipette 5 mL aliquot from the digested plant tissue solution into 25-mL volumetric flask and add 1 mL of the EDTA solution. After thoroughly mixing, allow the flask to stand for 1 min. Now add 2 mL of phenol-nitroprusside reagent and immediately make up the volume with distilled water. Mix the solution thoroughly and place the flask in water-bath for 30 min maintained at 40^0 C temperature. Remove the flask from water-bath and cool at room temperature at least for

10 min. Determine the absorbance of the developed colour complex at a wavelength of 636 nm against a reagent blank solution. Determine the NH_4-N concentration of sample by reference to a calibration curve.

Preparation of standard curve

From stock standard solution of 50 μg/ ml NH_4-N, prepare working solution of 2 μg/ml NH_4-N. Take the same volume of distilled water as was used in aliquot of plant extract) to a series of 25-ml volumetric flasks. Add 0, 1, 2, 3, 4, 5 and 6 mL of the working solution i.e. 2 μg NH_4-N /ml to obtained 0, 2, 4, 6, 8, 10 and 12 μg concentration of NH_4-N. Develop the colur with standards by adopting the procedure as described for NH_4-N analysis in plant solution. Measure the colour intensity at 636 nm wavelengths. Draw a curve between absorbance and concentration of NH_4-N in solution on a semi-log paper.

Calculations

NH_4-N (ppm) = Test solution-N (from curve) x Aliquot Dilution x Tissue dilution

2.5.2. *Nesslerization method* (Kanwar and Chopra, 1976)

Principle

When Nessler's reagent is added to a known volume of dilute plant digested solution or distillate containing ammonium, the ammonia is liberated and reacts with the reagent fairly rapidly but not instantaneously, to produce colloidal orange-brown product whose intensity is measured on a spectrophotometer at 420 nm wavelength and compared with the reading of standard NH_4-N solution from the curve.

Reaction

$$2\ K_2(HgI_4) + 2\ NH_3 \rightarrow NH_2Hg_2I_3 + 4\ KI + NH_4I$$

Reagents

(i). **Nessler's reagent**: Dissolve 45.5 g of mercuric iodide and 35.5 g of potassium iodide in about 250 mL hot distilled water in a 500-mL beaker and transfer it to a 1000-mL volumetric flask. In another beaker, dissolve 112 g of KOH in about 400 mL of distilled water. Mix this solution with iodide solution by giving repeated washing with distilled water. The solution is mixed thoroughly, cooled and diluted to complete one litre volume with distilled water.

Allow this solution to stand for 3-4 days and thereafter, the clear supernatant liquid is decanted off into a amber colored bottle for use.

(ii). **Sodium silicate (10%) solution**: Add 10 g of sodium silicate in about 50-60 mL of distilled water. Warm it to dissolve completely and cool. Make complete volume of 100 mL with distilled water.

(iii). **Standard NH_4-N (50 ppm) solution**: Prepare as described in section 2.1 of this chapter.

Procedure

Take 0.2 to 1.0 mL (depending upon N concentration) of the dilute digested aliquot directly or distillate of aliquot (obtained with MgO or NaOH distillation) into 25 mL volumetric flask. Dilute it to about 10 ml by adding distilled water. Add 1 mL of sodium silicate in it to neutralize the acidity. Make about 20 mL volume by performing neck washing of flask with distilled water. Add 2 mL of Nessler's reagent to it and mix thoroughly. Make complete 25-mL volume with distilled water, shake well again and allow standing for about 10-15 min to develop colour. Measure the intensity of the colour in terms of transmittance/ absorbance on spectrophotometer at 410 nm wavelength. Compare the transmittance/ absorbance of the test solution with that of the standard solution from the curve, prepared by taking series of standard NH_4-N solutions.

Preparation of standard curve

From the stock standard solution of 50 μg/mL NH_4-N, prepare working solution of 2 μg/ mL NH_4-N. Take appropriate volumes of distilled water (same volumes as used for aliquot of plant digest) in a series of 25-mL volumetric flasks. Add 0, 1, 2, 3, 4, 5 and 6 ml of the 2 μg/mL NH_4-N working solution to obtained 0, 2, 4, 6, 8, 10 and 12 μg concentration of NH_4-N. Develop the colour with standards by adopting the procedure as described for analysis of NH_4-N in digested plant aliquot. Measure the colour intensity either in terms of transmittance/ or absorbance at 420 nm wavelengths. Plot a graph between transmittance/ or absorbance versus concentration of NH_4-N (ppm) in solution on semi-log paper. If the concentration obeys Beer's Law, then a linear calibration curve is obtained. From this graph, calculate the concentration of the given unknown solution.

Calculations

NH_4-N (ppm) = Test solution-N (from curve) x Aliquot Dilution x Tissue dilution

2.6. Salicylate- hypochlorite method (Crooke and Simpson, 1971)

Since phenol is known today as a harmful carcinogenic compound, hence, this reagent should be replaced by less toxic compounds, such as salicylate- hypochlorite method.

Principle

The reaction of NH_4 ions with a weak alkaline mixture of sodium salicylate and chlorine gives a yellow color. Sodium nitroprusside solution is added to speed up the colour reaction at room temperature. The colour intensity is measured with help of spectrophotometer.

Apparatus: Spectrophotometer

Reagents

(i). **Sodium salicylate-sodium nitroprusside solution**: Dissolve 300 mg of sodium nitroprusside [$Na_2Fe(CN)_5NO.2H_20$] and 150 g sodium salicylate [$NaC_7H_5O_3$] in 600 mL distilled water.

(ii). **Sodium hypochlorite (NaOCI or Clorox)**: Dissolve 6 mL Clorox containing 5.25% Cl in 100 mL distilled water.

(iii). **Buffer solution**: Dissolve 50 g of sodium-potassium tartrate ($NaKC_4H_4O_6.H_2O$) and 26.8 gm of di-sodium phosphate ($NaHPO_4$) in about 600 mL of distilled water. Dissolve 54 gm of NaOH into it by constant stirring. Make complete one liter volume with distilled water.

(iv). **Standard NH_4-N (50ppm) solution**: Prepared as described in section 2.1 of this chapter.

Procedure

Pipette out 1 mL of the diluted Kjeldahl digested solution into a colorimetric tube or cuvette. Add to it 5 mL of the buffer solution and 4 ml of sodium salicylate-nitroprusside solution and stir. Now add 2 ml of sodium hypochlorite solution and again stir to mix the mixture thoroughly. Allow the solution to stand in water-bath for 45 min at 25°C for the development of yellow colour. Remove the colorimetric tube from water-bath and cool at room temperature for 10 min. Measure the absorbance of the coloured solution at 650 nm wavelength. Compare N concentration of plant aliquot with the standard curve prepared by taking readings of series of standard working solutions of NH_4.

Calculations

NH_4-N (ppm) = Test solution-N (from curve) x Aliquot Dilution x Tissue dilution

2.7. Ammonium determination method (Yuen and Pollard, 1952)

Principle

When a N-containing solution is treated with Sodium-potassiumtitrate it gives yellow color in presence of sodium potassium mercuric iodide. The intensity of colour is measured colorimetricaly

Reagents

(*i*). *Sodium-Potassium-Mercuric iodide reagent*: Dissolve 22.72 g of mercuric iodide in approximately 50 mL of water solution containing 18.26 g of potassium iodide. This solution is slowly added with constant stirring to about 900 mL solution containing 40 g of NaOH. Cool and make complete one litre volume with distilled water.

(ii). **Sodium-potassium-tartrate solution (10%)**: Dissolve 50 g of sodium potassium tartrate in 500 distilled water and add 2-3 mL of chloroform in it.

(iii). **Standard NH_4-N (50ppm):** Prepare as described in section 2.1 of this chapter

Procedure

Take 10 mL of digested or distillate plant solution in 50-mL volumetric flask. Dilute it to about 30 mL with distilled water. Add 10 mL of sodium potassium tartrate solution with subsequent shaking. The flask is placed in a water bath at 25 ± 0.5°C until its contents reach to bath temperature. Now add 5 mL of sodium-potassium mercuric iodide reagent and make up the volume with distilled water and place the flask over the water-bath again. After 20-30 minutes yellow color will appear. Determine the color intensity of solution on colorimeter at 435 nm. Compare the transmittance percentage of solution with the standard curve prepared with standard NH_4-N solution of different concentration following the earlier mentioned procedure.

Calculations

NH_4-N (ppm) = Test solution-N (from curve) x Aliquot Dilution x Tissue dilution

2.8. Specific ion electrode methods (Bremner and Tabatabai, 1972)

The determination of inorganic forms of nitrogen by specific electrodes would be quite attractive when compared with manual distillation or colorimetric methods. Before adaptation of specific ion electrode method, specific ion electrodes should be carefully evaluated for their suitability for a particular application. Bremner and Tabatabai (1972) first reported the use of NH_3 electrode for analysis of NH_4^+ in Kjeldahl digest and their results agreed closely to conventional alkaline distillation and titration method. In this method, distillation step could be omitted.

Principle

Ammonium-N concentration in a test solution is estimated by comparison of electromotive force (emf in milli-volts) in the unknown solution with that obtained by analysis of NH_4^+-N standards by the same method. The sample or standard is obtained by the addition of 0.2M NaOH solution (pH 11-12), because the electrode responds only to NH_3 activity.

Apparatus: 1. Ammonia electrode

2. pH-millivolt meter with a sensitivity of $\pm$ 0.1 mV

Reagents

1. **Sodium hydroxide (NaOH) 0.25 M**: Dissolve 10 g of NaOH in one liter of distilled water.
2. **Standard NH_4 -N solutions**: Prepare a series of standards in distilled water from 0.1 to 10 ug of NH4+-N/ mL.

Procedure

Take 20 mL aliquot of digested plant sample in a 50-mL beaker containing a Teflon-coated stirring bar. Place the beaker on a magnetic stirrer, add 2 mL of 0.25 M NaOH solution. Insert a NH_3 electrode connected to a pH-millivolt meter into the solution. Stir the solution for 1 min, record the electrode potential value and calculate NH_4^+-N in the sample by comparison of this value to a calibration curve prepared as described for determination of NH_4^+-N. Observations must be recorded within 1-2 min. after the addition of NaOH to ensure no loss of NH_3. Insert the electrode at about 20^0 angle with respect to vertical axis.

Calculations

$$NH_4\text{-}N\ (ppm) = \text{Test solution-N (from curve) x Aliquot Dilution x Tissue dilution}$$

REFERENCES

Bremner, J.M. (1960). Determination of nitrogen in soil by the Kjeldahl method. *Journal of Agricultural Science* **55:** 11-33.

Bremner, J.M. (1965). Inorganic forms of nitrogen. In: *Method of Soil Analysis. Part 2.* American Society of Agronomy, Madison, WI, USA. *Agronomy Monograph* **9**: 1179-1237.

Bremner, J.M. and Keeney, D.R. (1965). Steam distillation methods for determination of ammonium, nitrate and nitrite. *Analytica Chimica Acta* **32**: 485-495.

Bremner, J.M. and Shaw, K. (1955). Determination of ammonia and nitrate in soil. *Journal of Agricultural Science* **51**:320-328.

Bremner, J.M. and Tabatabai, M.A. 1972. Use of an electrode for determination of ammonium in Kjeldahl analysis of soils. *Communications in Soil Science and Plant Analysis* **3:** 159-164.

Conway, E.J. (1947). *Micro Diffusion Analysis and Volumetric Error.* 2nd edn., Crosby Lockwood, London.

Crooke, W.M. and Simpson, W.E. (1971). Determination of ammonia on Kjeldhal digest of crops by an automated procedure. *Journal of Science, Food and Agriculture* **22**: 9-10.

Dumas, J.B.A. (1831). *Ann. Chim. Phys.* (Paris), (2), 47, 198.

Kanwar, J.S. and Chopra, S.L. (1976). Available nutrients and exchangeable cations in soils In: *Analytical Agricultural Chemistry.* Kalyani Publishers, New Delhi, pp. 255.

Kempers, A.J. (1974). Ditermination of sub-microquantites of ammoniaum and nitrates in soils with phenol, sodium with nitroprusside and hypochlorite. *Geoderma* **12** : 201-206.

Kjeldahl, J. (1883). *Z. Anal. Chem.* **22:** 366.

Yuen, S.H. and Pollard, A.G. (1952). The determination of nitrogen in agricultural materials by the Nessler reagent. 1. Preparation of reagent. *Journal of Science, Food and Agriculture* **3**:441-447.

Research Methods in Plant Sciences : Allelopathy.
Volume 4. Plant Analysis
Eds. S.S. Narwal, O.P. Sangwan and O.P. Dhankhar
Scientific Publishers (India), Jodhpur (2007) (pp. 87-96)

7

PHOSPHORUS

Tek Chand and D. S. Mehla

Department of Soil Science, CCS Haryana Agricultural University,
Hisar-125 004, India

1. INTRODUCTION

Next to nitrogen, phosphorus is major element vital for plant growth and development. It is structural component of metabolically active compounds in plants. Its concentration and solubility in soils is very low due to its interaction with soil colloids. The total P content of Indian soils is 0.03 to 0.25% but the concentration of available P forms varies from 0.005 to 0.03% of total P. In India, Phosphorus deficiency has become widespread in more than 98% of soils due to intensive cropping. Among several diagnostic methods available, analysis of growing plants is most efficient method. Analysis of plant materials, shows that whether the nutrient is in deficiency, optimum or excessive range. The concentration of P in different plants under low, sufficient and high ranges has been determined (Bhargava and Raghupathi, 1993) and it varies from 0.05 to 1.5 percent.

2. DETERMINATION OF PHOSPHORUS

To determine phosphorus, plant samples may be digested in a diacid (HNO_3–$HClO_4$ or H_2SO_4–$HClO_4$) or triacid (HNO_3–H_2SO_4–$HClO_4$) mixture or dry ashed and dissolved in acid (See Chapter.....).

The plant sample digests are analyzed for their elemental composition using one or more techniques : Colorimetry, Flame Emission, Atomic Absorption, Spark/Arc Emission, Inductively Coupled Plasma Atomic Emission Spectrometry (ICP-AES) and Auto Analyzer, etc. The most important methods are described in the following sections.

2.1. Vanado-molybdophosphoric Yellow Colour Method (Koenig and Johnson, 1942, Modified by Jackson, 1958)

Generally the analysis of P in plant extracts is done using low sensitive methods e.g. vanado-molybdophosphoric yellow colour method in nitric acid.

Principle

Plant phosphorus is changed into orthophosphates during the digestion. These orthophophates react with molybdate and vanadate and give yellow coloured vanado-molybdophosphoric heteropoly compounds in acid medium. The yellow colour is attributed to substitution of oxyvanadium and oxymolybdenum radicals for the O (oxygen) of PO_4. The intensity of this colour is directly proportional to the concentration of phosphates present in the sample, which can be measured on spectrophotometer.

Apparatus

Weighing Balance, 100 ml digestion flasks, 25 mL volumetric flasks, pipettes, wash bottles, spectrophotometer.

Reagents

i. **Ammonium molybdate-vanadate :** Dissolve 25 g ammonium molybdate $[(NH_4)_6 Mo_7 O_{24}. 4H_2O]$ in 400 mL distilled water. Dissolve 1.25 g ammonium vanadate in 300 mL boiling distilled water. Add the vanadate solution to the molybdate solution and cool to room temperature. Add 250 mL concentrated HNO_3 along with intermittent stirring. Make up the volume to 1 litre with distilled water.

ii. **2,4 dinitrophenol indicator :** Weigh 0.2 g 2,4 dinitrophenol and dissolve in 100 mL ethanol.

iii. **6N HCl :** Add 250 mL concentrated hydrochloric acid (HCl) to a 500 mL volumetric flask containing 250 mL distilled water and mix well.

iv. **Standard phosphate solution:** Dissolve 0.2195 g of potassium dihydrogen phosphate (KH_2PO_4) in 400 mL of distilled water. Add 25 mL of 7N H_2SO_4 and make up the volume to 1 litre with distilled water. This will give 50 ppm P solution.

Procedure

Take 5 mL plant digest aliquot (0.5 g plant material in 50 mL) in 25 mL volumetric flask. Add 2-3 drops of 2,4 dinitrophenol indicator. After neck washing with distilled water, add ammonia solution drop by drop till the yellow colour appears. Then, add 6N HCl drop by drop till the yellow colour disappears. Then, add 5 mL of vanadomolybdate reagent, make up the volume to 25 mL with distilled water and shake thoroughly. The colour develops in about 30 minutes and is stable for pretty long time. Read the transmittance or absorbance of the solution after 30 minutes at 440 nm on spectrophotometer/colorimeter using blue filter. Record P concentration in solution from standard curve and compute/calculate P concentration in plants employing dilution factor.

Preparation of standard curve : Prepare the standard curve by taking the aliquots containing 0.5 to 10 mg L^{-1} (ppm) of P in 25 mL volumetric flasks and develop the colour as done for plant digest. Read the transmittance or absorbance of the solution after 30 minutes on spectrophotometer at 440 nm using blue filter. Plot the transmittance curve against concentration.

Calculations

Weight of plant sample	=	0.5 g
Volume of digest	=	50 mL
Aliquot taken	=	5 mL
Conceneraction. of P in aliquot	=	C ppm

$$P\,(\%) = \frac{C}{\text{Wt of sample (g)}} \times \frac{100}{\text{Aliquot taken (mL)}} \times \frac{\text{Volume of Digest (mL)}}{10000}$$

$$= \frac{C}{0.5} \times \frac{100}{5} \times \frac{50}{10000} = \frac{C}{5} = 0.2C$$

Advantages

The main advantages of this method are as under : (i) very simple, (ii) low sensitivity (about 1/10 of blue colour method), (iii) stability of colour (iv) freedom from interferences by many ionic species in concentrations upto 1000 ppm.

2.2. Autoanalyzer method (Thomas *et al.*, 1967)

Principle

Plant phosphorus is changed into orthophosphates during the digestion with H_2SO_4– H_2O_2. The estimation of orthophosphates is based on a colorimetric method, in which the blue colour is formed by the reaction of molybdate and orthophosphate ions followed by reduction with ascorbic acid. This reaction is very sensitive to pH.

Apparatus

i. Standard units of the autoanalyzer system is used in the arrangement illustrated in the flow diagram (Fig.1) for the combined N and P analysis.

ii. Tube sampler equipped with a special arm capable of moving to the side to pump distilled water between samples.

iii. Two colorimeters and recorder if they are not available, two separate manifolds can be constructed to measure P or N independently.

Reagents

i. **Ammonium molybdate [$(NH_4)_6$ Mo_7 O_{24} $.4H_2O$] :** Dissolve 20 g ammonium molybdate in 400 ml hot distilled water and cool. Add 225 mL concentrated H_2SO_4 to 400 mL distilled water and cool. Add the molybdate solution to the H_2SO_4 solution with continuous stirring and make the volume to 2.0 litre with distilled water.

ii. **Ascorbic acid:** Dissolve 1.76 g ascorbic acid in 100 mL distilled water and make volume to 200 mL. A fresh solution should be prepared every time for determination.

iii. **0.36 M H_2SO_4:** Add 20 mL concentrated H_2SO_4 to 900 mL and make final volume to 1 litre.

iv. **P-Diluent:** Add 1 mL Levor V to 1 litre distilled water.

v. **H_2SO_4 concentrated**

vi. **H_2O_2 (30%)**

vii. **Standard phosphate solution:** Dissolve 0.4395 g potassium dihydrogen phosphate (KH_2PO_4) in distilled water and make volume to 1 litre with distilled water. It is 100 ppm stock solution. Working standards containing P from 0-10 mg P L^{-1} are prepared in 0.36 M H_2SO_4.

Procedure

1. Weigh 0.25 g plant sample in a 100 ml digestion flask. Add 5 ml concentrated H_2SO_4 and heat on a hot plate at 225°C until the condensation line of H_2SO_4 left 1.0 cm above the bottom of the flask. After cooling, add four drops of 30% H_2O_2 and heat again on the hot plate for 2 to 5 min. Repeat this process at least for 12 times. After cooling, dilute the digest to 250 ml with distilled water for containing P concentrations 0.1% to 1.2% in plant materials.
2. Set up the auto-analyzer as shown in Fig. 1 and place the plant digest in tube sampler of the auto-analyzer. Automatically the air, ammonium molybdate solution and the ascorbic acid are sucked and mixed in the required volumes at calibrated fixed rates.
3. The sample is then heated for approximately 4.0 minutes at 95°C in a coil provided in the oil bath and finally passed through a 15 mm tubular flow curvette after separating air bubbles to waste pipe.
4. The colour intensity is measured at 815 nm on the colorimeter and peaks are recorded.
5. The values are compared with the standard curve.
6. The standard curve is prepared by using the standards ranging from 0-10 mg P L^{-1} in 0.36 M H_2SO_4 and developing the colour as described above for plant samples and measuring the colour intensities of the solutions on colorimeter and recording the peaks on the recorder.

Calculations

Weight of Plant sample = 0.25 g

Final volume of digest = 250 mL

Dilution factor = 1000 times

P (ppm) = autoanalyzer reading mg L^{-1} x dilution factor = x

P (%) = x/10000

Comments

This reaction is very sensitive to pH, some adjustments may be required to the ammonium molybdate solution if a dilution loop is attached to the manifold. Phosphorus can also be determined in extracts of dry ashing diacid digest of HNO_3-$HClO_4$ with, some adjustments to the ammonium molybdate solution. Stannous chloride can be substituted to ascorbic acid with the result that the method is less sensitive to pH fluctuations.

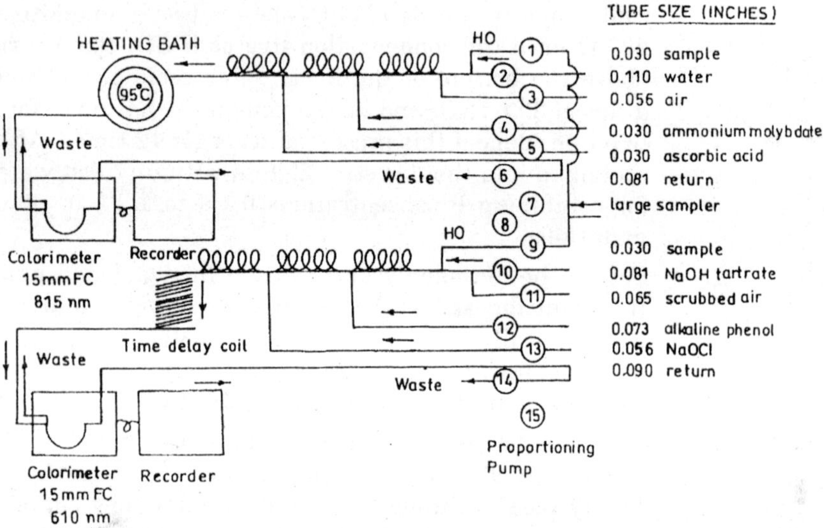

Fig. 1. A flow diagram of autoanalyzer manifold for simultaneous determination of N and P

The major advantage of the automated system of analysis for N and P, is the ease of adjusting for samples which do not fall within the normal operating range. Alterations of sample size, dilution tube size or both permit divergent groups of samples to be re-run with a new set of standards in minimum time. The combined N and P analysis in the same digest saves time. For greater precision, duplicate and triplicate determinations may be done in with less time than conventional analysis.

2.3. Inductively Coupled Plasma - Atomic Emission Spectrometry method (Isaac and Johnson, 1985)

Plasma emission spectroscopy is useful for multi-elements determination in plant tissue. The plasma system is much more adaptable to matrix changes and sample types as dynamic linear range enables the analysis of low and high concentrations of the same element.

Principle

Sample is dry-ashed, treated with HNO_3 and dissolved in HCl; P is detected by Inductively Coupled Plasma - Atomic Emission

Spectrometry (ICP-AES). The ICP source produces a stream of high energy ionized gas by inductively coupling an inert gas (commonly argon) with a high frequency field. When a fine mist of liquid sample is introduced into the centre of the plasma, it desolvates, dissociates, atomizes and excites the elements in the sample. When the generated ions return to their atomic ground state, light (photons) is emitted, the intensity of light measuring the concentratrion of element and wave length identify the element.

Apparatus

Inductively coupled plasma emission spectrometer, 30 mL glazed porcelain crucible, 50 ml volumetric flasks, muffle furnace, hot plate, source of argon (Ar) gas (liquid is preferred).

Reagents

(i) **Nitric acid (NO_3), 8 N :** Add 250 mL concentrated nitric acid (HNO_3) to a 500 mL volumetric flask containing 250 mL distilled water and mix well.

(ii) **Hydrochloric acid (HCl), 6 N:** Add 250 mL concentrated hydrochloric acid (HCl) to a 500 mL volumetric flask containing 250 mL distilled water and mix well.

(iii) **Stock solution of P (1000 ppm):** Weigh 3.7138 g of ammonium phosphate ($NH_4H_2PO_4$) in 1.0 litre volumetric flask, dissolve in 400 mL distilled water and make the volume with distilled water.

(iv) **Standard phosphate solutions (5 ppm, 60 ppm):** Pipet 5 mL and 60 mL volumes of stock solutions into 1litre volumetric flasks. Add 100 mL HCl and make volume to 1.0 litre with distilled water.

Procedure

Weigh 1 g plant sample into a 30 mL glazed porcelain crucible. Place the crucible on a muffle furnace at 500° C for 2 hours. Remove the crucible from muffle furnace, cool and add 3 mL HNO_3. Heat the sample on a hot plate at 100 to 120° C to evaporate excess HNO_3. Place the crucible back to the muffle furnace and muffle at 500° C for additional hour. Cool the crucible, add 10 mL HCl and transfer the sample to a 50 mL volumetric flask. Dilute to volume with distilled water and mix well.

Analyze the obtained plant tissue ash solution as per AOAC Method 985.01 (Helrich, 1995) or follow the recommendations of instrument manufacturer. Phosphorus is determined by inductively coupled plasma emission spectroscopy. Calibration of instrument is done using the known calibration standards. The Suggested operating parameters for ICP emission spectrometer are : forward power 1.1 kilowatts, reflected power <10 watts, aspiration rate 0.85-3.5 ml/min., flush between samples 15-45 s, integration time 1-10 s and wave length 214.9 x 2[a] nm ([a] second order position) or 2149Å. After calibration is complete, sample can be analyzed. Check calibration after every 10 samples. If instrument has drifted out of calibration (>3% of original values), recalibrate the instrument.

Calculations

Concentration for P of diluted sample is determined as μg/mL.

Advantages

The advantages of ICP-AES are: (i) Minimum chemical interferences, (ii) Rapid, accurate and precise analysis, (iii) Multi-elements determination capabilities and (iv) Detection limits equal to or better than atomic absorption spectrometry (AAS).

Disadvantages : (i) Occurrence of special interferences and (ii) Use of argon gas is expensive.

Common Methods : The common methods for elemental and ion determination in plant extracts, digests and ash solutions based on their solubility are listed in Table 1 (Watson, 1998).

Table 1. Instrumental methods for analysis of plant nutrients

Element	Method of Analysis						
	Colori-metric	Emission		ICP	X-ray	Atomic absorption	Specific ion electrode
		Flame	Spark				
Phosphorus (P)	Ex	NA	Ex	Ex	Fair	NA	NA
Boron (B)	Good	NA	Good	Ex	Poor	NA	NA
Calcium (Ca)	Good	Fair	Good	Ex	Poor	Ex	Poor
Copper (Cu)	Good	NA	Good	Ex	Ex	Ex	NA
Iron (Fe)	Fair	NA	Good	Ex	NA	Ex	NA
Magnesium (Mg)	Fair	Fair	Ex	Ex	Poor	Ex	NA
Manganese (Mn)	Good	NA	Ex	EX	Poor	Ex	NA
Molybdenum (Mo)	Good	NA	Poor	Good	Good	Good[a]	NA
Potassium (K)	Poor[b]	Ex	Ex	Ex	NA	Good	NA
Sodium (Na)	NA	Ex	Ex	Ex	NA	Good	Fair
Zinc (Zn)	Good	NA	Ex	Ex	Good	Ex	NA
Nitrate (NO_3)	Ex	NA	NA	NA	NA	NA	Good
Ammonium	Good	NA	NA	NA	NA	NA	Good

Chloride	Good	NA	NA	NA	NA	NA	Good
Fluoride	NA	NA	NA	NA	NA	NA	Good
Sulfate	Good[b]	NA	NA	NA	Good	NA	NA

NA = Not applicable; Ex = Excellent (high sensitivity with minimal interference); Good = Moderate sensitivity with some interference; Fair = Reasonable sensitivity but with matrix effects; Poor = Reasonable sensitivity with significant matrix effects.

[a] Turbidity, [b] Flameless AA

3. REFERENCES

Bhargava, B.S. and Raghupathi, H.B. (1993). Analysis of plant materials for macro and micronutrients pp 49-82. In: *Methods of Analysis of Soil, Plant, Water and Fertilizers* (Ed., H.L.S. Tandon). Fertilizers Development and Consultation Organisation, New Delhi.

Helrich, K. (Ed.). (1995). *Official Methods of Analysis of the Association of Official Analytical* Chemists. Volume 1. Method 985.01. Association of Official Analytical Chemists, Arlington, VA., USA.

Isaac, R.A. and Johnson, Jr., W.C. (1985). Elemental analysis of tissue by plasma emission spectroscopy : Collaborative study. *Journal Association Office Analysis Chemistry* **68:** 499-505.

Jackson, M.L. (1958). *Soil Chemical Analysis*. Prentice- Hall, New Delhi.

Koenig, R.A. and Johnson, C.R. (1942). Colorimetric determination of phosphorus in biological materials. *Industrial and Engineering Chemistry* **16 :** 233-237.

Watson, M.E. (1998). Analytical instruments for the determination of elements in plant tissue. In: *Handbook of Reference Methods for Plant Analysis*. pp 137-152. Soil and Plant Analysis Council, Inc. Washington, D.C.

Thomas, R.L., Sheard, R.W. and Moyer, J.R. (1967). Comparison of conventional and automated procedures for nitrogen, phosphorus and potassium analyses of plant tissue using a single digestion. *Agronomy Journal* **59 :** 240-243.

Research Methods in Plant Sciences : Allelopathy.
Volume 4. Plant Analysis
Eds. S.S. Narwal, O.P. Sangwan and O.P. Dhankhar
Scientific Publishers (India), Jodhpur (2007) (pp. 97-104)

8

POTASSIUM AND SODIUM

S.C. Mehta and K.S. Grewal

Department of Soil Science, CCS Haryana Agricultural University, Hisar-125 004, India

1. INTRODUCTION

Potassium plays an important role in the growth and development of plants. It (i) promotes photosynthesis, (ii) helps in formation of carbohydrates, oils, fats and proteins, (iii) activates more than 60 enzymes, (iv) and promotes the transport of photosynthates to seeds, roots, fruits and tubers, (v) increases the ability of plants to withstand drought and frost, lodging and attack of pests and diseases, (vi) increases the quality of crops and (vii) helps in better utilization of other plant nutrients like nitrogen, phosphorous and sulphur.

Sodium is present in insoluble silicate minerals and total Na in soils is generally less than 2%. The water-soluble sodium also varies from traces to very high amount depending upon soil salinity. Sodium (i) helps in maintaining the turgor pressure in the cell sap (ii) controls the Na: K ratio in the plant. Tree crops and woody perennial plants are sensitive to low concentration of sodium, but annual crops are affected by its higher concentrations. Leaf burn, scorch and dead tissues along the outside edges of the leaves are typical toxicity symptoms. For useful plant analysis, it is prerequisite that the plant part should be properly selected, collected, processed and analyzed (Kensworthy, 1964; Reuter and Robinson, 1986; Jones *et. al.*, 1991; Bhargava and Raghupathi, 1993; Singh *et al.*, 1999). Hence, efforts should be made to select proper plant part, stage of growth and sampling time for plant analysis. The details regarding the collection and preparation of different plant parts (leaf, shoot, roots, grains etc.) have been explained in Chapter 3 of this book.

2. POTASSIUM AND SODIUM IN PLANTS

Potassium may be determined directly, in the solution of plant ash, either by the perchlorate or the volumetric cobaltinitrite methods. Sodium can be determined in the solution of plant ash by precipitation as sodium uranyl magnesium acetate [Na $(UO_2)_3$ Mg $(CH_3COO)_9 . 8H_2O$]. These methods are described by Piper, (1966) but are crude, tedious and time consuming, thus not used and their details are not given in this chapter. Potassium and sodium in the plant samples are usually determined by digesting the plant in di-acid mixtures (wet oxidation) see Chapter 4. Then K and Na in acid digest is determined using Emission Flame Photometric Method.

2.1. Emission Flame photometric Method

Principle

It is based on the principle that atoms of some specific elements absorb energy from the flame and get excited to the higher orbit. Such atoms release energy of a characteristic wavelength, which is specific for those elements and is proportional to the concentration of atoms of those elements in the sample.

To estimate the potassium and sodium, the instrument known as "Flame Photometer" is generally used. It works on the principle of flame photometry. If a solution containing metallic salt (or some other metallic compound) is aspirated into a flame, a vapour that contains atoms of the metal may be formed. Some of these gaseous metal atoms may be raised to an energy level, which is sufficiently high to permit the emission of radiations characteristic of the metal when the atoms drop down to less energetic states or to the ground state. Excited atoms and ions give "line" radiations at very definite wavelengths. The most prominent lines of sodium and potassium and the radiations most usable for flame analysis in soils and plants works are the doublets at 589.0 and 589.6 mμ for Na and 766.0 and 769.0 mμ for K.

Equipment and apparatus

Flame photometer, funnels, beakers, pipette, conical flasks, volumetric flasks, and filter papers.

Types of flame photometers

Flame photometers are of two types - single beam and double beam.

(i) ***Single beam type***

Single beam equipment commonly referred to as direct reading type, comprises only one set of optics. Light emitted from the core of the flame just above the inner cone is collected by reflector and focussed by a lens of heat resistant glass through interchangeable optical filters (interference type filter) on to a single photo-detector. Alternatively, light from the burner passes into monochromator and radiation leaving the exit slit is focussed on photo-detector unit.

(ii) ***Double beam type***

In the double beam type instrument, a second light path is provided for the radiation emitted by the internal standard element (e.g. lithium) that is added in a fixed amount to each test solution as well as calibration standard. The wavelengths of internal standard and analytical materials are isolated by special filters in a dual optical system and each beam is focussed on separate photoelectric cell.

Single beam type flame photometers are in common use and their operation is described below:

Operation

The general principle of operation of a flame photometer is depicted in Fig. 1. Although the ordinary type of manual flame photometer can be operated with great ease, but the modern flame photometers are operated by computer. The actual steps involved may vary with the type of instrument. However, the common steps required to estimate Na and K are given as under :

Reagents

- **Standard K solution**: Dissolve 1.908 g dried potassium chloride (KCl) salt in distilled water in one litre volumetric flask and make up the volume to 1 litre with distilled water. This solution will be the stock solution of 1000 ppm K. From the stock solution, make the working solutions of 5, 10, 15, 20, 25, 30, 35, and 40 ppm K by dilution.

- **Standard Na solution**: Dissolve 2.543g of dried sodium chloride (NaCl) salt in distilled water in one litre volumetric flask and make up the volume to 1 litre with distilled water. This solution will be the stock solution of 1000 ppm Na. From the stock solution, make the working solutions of 5, 10, 15, 20, 25, 30, 35, and 40 ppm Na by dilution.

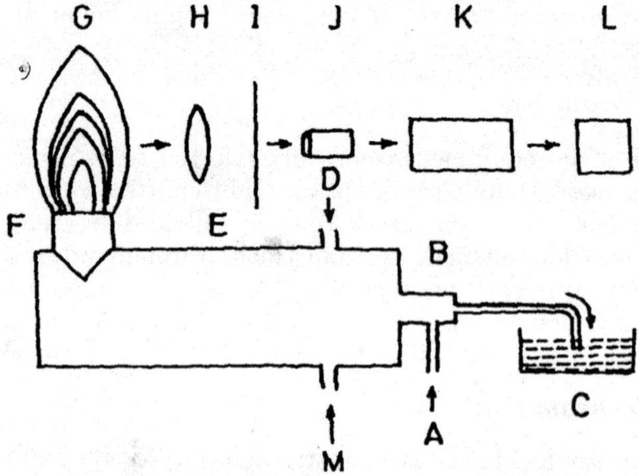

Fig. 1. Principles of operation of Flame photometer
A- air, B - atomizer, C - solution/sample, D - gas, E - mixing chamber, F - burner, G - Flame, H - lens system, I - filter, J - detector, K - amplifier and signal processor, L - read out, M - drain

Procedure

A. Calibration of flame photometer

The following steps are used for calibration of flame photometer:

i. Select the desired wavelength or optical filter
ii. Switch on the instrument.
iii. Switch on the compressor to get air supply.
iv. Adjust the air control to give a pressure of 0.75 kg/cm^2 (10 PSI).
v. Leave the instrument on for about 5 minutes to get it stabilized.
vi. Turn on the gas supply and light the flame.
vii. Adjust the gas control until formation of separate blue cones in the flame.
viii. Set the galvanometer to zero by control knob against blank solution.

ix. Set the galvanometer to 100 with the highest concentration standard solution.

x. Repeat steps no. 8 and 9 till the above setting is obtained consistently.

xi. Proceed with the determination of the standards and samples.

xii. For switching off the instrument, close the gas supply first, run distilled water for about 2 minutes and then switch off the compressor. Turn off the knobs and switches.

Precautions

i. The instrument should be connected to mains through a good voltage stabilizer.

ii. Keep the standards and samples free from any suspended material for preventing the clogging of the capillary. If the sample contains solid suspended particles, filtration should be done before analysis.

iii. Air pressure should be maintained at 0.75 kg/cm^2 (10 PSI).

iv. The gas inlet should be opened after opening the air inlet and closed before shutting of the air supply.

v. Do not feed highly concentrated acid and salt solutions.

vi. Do not leave the flame unattended.

vii. Add few drops of isobutyl alcohol to improve spray ability of the sample solution/extract.

viii. The standard curve should be drawn afresh each time the instrument is used.

ix. Periodically replace the pressure tubing of the fuel gas.

x. Keep the instrument protected from dust particles and acid fumes.

B. Preparation of Standard curve

i. Take the above mentioned reagents i.e. a series of standard solution of Na and K (0, 5, 10, 15, 20, 25, 30, 35, and 40 ppm).

ii. After calibration of the flame photometer, repeat calibration steps viii and ix once again. Feed all the standard solutions one by one starting from the lowest concentration and record reading of each concentration for K/Na.

iii. Draw the standard curve on a simple graph paper taking concentration on X-axis and reading of the flame photometer on Y-axis obtained against each concentration.

C. Measurement of Na or K concentration in test solution

The digested plant and /or grain samples can be tested for estimation of Na and K using flame photometer. Depending upon the concentrations of K and Na in the digests of plant samples, a digest can be used either directly or after dilution for flame photometric determinations.

Take the plant digest (either directly or after dilution) and feed to the flame photometer and record the reading in the same way as explained in section B above. Let the reading of the test solution be x.

Calculations

Calculate Na or K content in sample, using the standard curve.

Weight of the plant material taken	= 0.5g
Volume made after digestion	= 50 ml
Dilution factor	= 50/0.5 = 100 times
Flame photometer reading (say)	= x
Na or K concentration (ppm) from standard curve against reading (x)	= Y
Total Na or K (ppm) in plant sample	= Yx100 dilution factor
Total Na or K (%) in plant sample	= Yx100/10000 = 0.01Y

Interpretation of Results

There is no general categorization of potassium and sodium in crops based on its concentration as it depends on type of crops, stage of sampling and nature of crops (salt tolerance etc.). The K status of some important crops based on K concentration in the plant tissue is given in Table 1.

Table 1. Potassium status of some field crops based on K concentration in plant tissue

Crop	Optimum growth stage for sampling	Recommended plant part for sampling	Crop K status [based on K concentration (%) in plant part]			
			Sufficient	Semi-deficient	Mild-deficient	Acute-deficient
Rice	Tillering (45d)	Top two leaves	2.3-1.8	1.8-1.5	1.5-1.2	<1.2
	Anthesis (65d)	Top two leaves	2.6-1.6	1.6-1.3	1.3-1.1	<1.1

Wheat	Tillering (45d)	All aerial parts	6.4-4.1	4.1-3.3	3.3-1.1	<1.1
	Anthesis (70d)	Top two leaves	5.1-3.2	3.2-2.4	2.4-1.3	<1.3
Sorghum	Vegetative (30d)	leaves	3.5-2.5	2.5-2.1	2.1-1.3	<1.5
Pearl Millet	Vegetative (30d)	leaves	4.5-2.9	2.9-2.1	2.1-1.3	<1.3
Mustard	Flowering (65d)	leaf blades	5.4-3.7	3.7-3.0	3.0-2.1	<2.1
Ground nut	Pod initiation(40d)	leaf blades	3.0-2.2	2.2-1.9	1.9-1.3	<1.3
Chickpea	Vegetative (30d)	leaf blades	1.6-1.3	1.3-1.1	1.1-0.9	<0.9
	Flowering (70d)	leaf blades	1.7-1.4	1.4-1.3	1.3-1.1	<1.1
Black gram	Flowering (40d)	leaf blades	5.3-3.6	3.6-3.0	3.0-1.8	<1.8
Potato	Tuber initiation (37d)	leaf blades	8.0-5.6	5.6-4.4	4.4-2.8	<2.8
Cotton	Vegetative (60d)	top leaf blades	3.5-2.5	2.5-2.2	2.2-1.4	<1.4
Sugarcane	Vegetative (90d)	2nd leaf blades	1.5-1.1	1.1-0.9	0.9-0.7	<0.7
		2nd leaf sheath	2.5-1.7	2.5-1.7	1.4-0.9	<0.9

Source: Rao and Sekhon (1989)

3. REFERENCES

Ayers, R.S. and Westcot, D.W. (1989). *Water Quality for Agriculture. Irrigation and Drainage Paper* **29**, FAO, Rome, p174.

Bhargava, B.S. and Raghupathi, H.B. (1993). Analysis of plant materials for macro and micro nutrients. In: *Methods of Analysis of Soil, Plant, Water and Fertilizers* (Ed., H.L.S. Tandon). Pp. 49-82. FDCO, New Delhi.

Dixon, J.B. and Weed, S.B. (1977). *Materials in Soil Environment.* Soil Science Society of America, Madison, WI, USA.

Hurlbut, C.S. (1971). *Dana's Manual of Mineralogy.* 18th Ed. John Wiley and Sons, Inc., New York.

Jones Jr. J.B., Wolf, B. and Mills, H.A. (1991). *Plant Analysis Handbook: A Practical Sampling, Preparation, Analysis and Interpretation Guide.* Micro-macro Publishing Inc. Athens, Georgia.

Kensworthy, A.L. (1964). *Fruit, Nut and Plantation Crops – A Guide for Collection of Foliar Samples for Nutrient Element Analysis.* Horticulture Department. Memeographed. Michigan State Univ. East Lansing, USA. pp.39.

Piper, C.S. (1966). *Soil and Plant Analysis.* Hans Publishers, Bombay, India

Rao, R.M. and Sekhon, G.S. (1989). *Critical Potassium Levels in Selected Crops.* Potash Research Institute of India, Gurgaon, India. Special Publication No. **2.** p. 67.

Reuter, D.J. and Robinson, J.B. (1986). *Plant Analysis- An Interpretation Manual.* Inkatta Press, Melbourne, Sydney.

Singh, D., Chhonkar, P.K. and Pandey, R.N. (1999). *Soil Plant Water Analysis: A Methods Manual.* Indian Agricultural Research Institute, New Delhi.

Research Methods in Plant Sciences : Allelopathy. Volume 4. Plant Analysis
Eds. S.S. Narwal, O.P. Sangwan and O.P. Dhankhar
Scientific Publishers (India), Jodhpur (2007) (pp. 105-118)

9

CALCIUM AND MAGNESIUM

O.P. Sangwan, A.K. Jain and V.K. Phogat

Department of Soil Science, CCS Haryana Agricultural University, Hisar-125004, India

1. INTRODUCTION

The bulk of total soil calcium and magnesium is present in structural combinations of soil minerals (calcite, magnesite, dolomite, olivine, apatite, feldspar, biotite and amphiboles) and their very small quantities exist in exchangeable and water soluble forms for uptake by plants. The nutrient uptake by plants depend on their availability in soils. The nutrients concentration in plants/plant parts reflects their status in soils where they have grown. Therefore, plant tissue analysis serves as a good tool to know available nutrients status in soil (Kensworthy, 1964; Tandon, 1984; Jones *et al.*, 1991).

Calcium is essential for the growth of meristems and root tips, whereas magnesium is an essential constituent of chlorophyll. Calcium and magnesium are also involved in activation of various enzymes reactions. Calcium and magnesium are absorbed by plants in the form of their ions as Ca^{2+} and Mg^{2+}, respectively. The concentration of Ca in calcium sufficient plants ranges between 0.2 and 1.0 per cent, whereas these values in respect of Mg range between 0.1 and 0.4 per cent. About 30% of the exchange complex of a soil must be saturated with calcium and magnesium for their sufficient uptake by plants of most crops. Calcium and magnesium are typically deficient in acid soils. In neutral soils, calcium and magnesium remains readily available to plants. Magnesium deficiency can be caused not only by small concentrations of the nutrient in a soil but also by ionic antagonism. An excessive concentration of calcium impairs the uptake of magnesium.

2. TOTAL CALCIUM AND MAGNESIUM

The digestion of plant material for Ca and Mg determination, is done with a di-acid mixture (concentrated HNO_3 and $HClO_4$) to bring the total plant Ca and Mg into solution/aliquot. The aliquot is used to determine of Ca and Mg concentrations. The details of the procedure have been described in Chapter 4 of this book.

There are two laboratory methods (Jackson, 1967) to determine Ca and Mg in a plant digest/aliquot : (i) Titrimetric method (ii) Emission Spectrophotometry method.

2.1. Titrimetric Method

This method is commonly used to determine Ca and Ca +Mg as an alternate method, owing to its speed and simplicity next to emission spectrophotometry (Jackson, 1967). During the titration of Ca or Ca plus Mg with EDTA, there can be the interference of heavy metals (Cu, Co, Fe, Mn, Ni, etc.) if they are present in appreciable quantities. This interference can be removed by using adequate amounts of NaCN and hydroxylamine hydrochloride, which will make stronger complexes with the heavy metals. Besides the pH of the titrating solution is raised to 12 for Ca and 10 for Mg by using NaOH and NH_4OH, respectively. The accuracy of this method is ± 3% (Hesse, 1971), which is considered reasonably well for its simplicity, therefore, commonly used for laboratory analyses.

2.1.1. *Determination of Calcium*

Principle

Calcium in plant digest (solution/aliquot) can be determined by titration of a known volume of the aliquot against the standard solution of EDTA using Calcon as indicator at pH 12 in the presence of NaOH. Adequate concentrations of NaCN and hydroxylamine hydrochloride are also used to remove the interference of heavy metals. The endpoint is the change of colour from pink to blue, when whole of Ca gets complexed with EDTA.

Apparatus

Magnetic stirrer, glass coated iron needles, ceramic china dishes of 100 mL capacity, wash bottles, micro burette, pipettes, beakers, volumetric flasks, etc.

Reagents

1. **Standard solution of Ca (0.01N):** Weigh 0.5 g dry $CaCO_3$ and transfer it to 400 mL beaker. Add 10 mL of 1N HCl in it and boil to expel CO_2. Cool it and add about 200 mL distilled water to dissolve the contents. Transfer it to 1 liter volumetric flask and finally make up the volume. Store in a glass reagent bottle.
2. **Standard solution of EDTA (0.01 N):** Dissolve 2.0 g EDTA-di-sodium salt in sufficient volume of distilled water taken in a 400 mL beaker. Transfer the contents to 1 liter volumetric flask. Add 0.05 g of $MgCl_2.6H_20$ and make up the volume to 1 liter. The normality of this solution is standardized by the following procedure (titration with standard Ca solution).

   Standardization of 0.01N EDTA solution: Pipette 5 mL of standard Calcium solution (0.01N) in a 100 mL China dish. Dilute it to 10 mL with distilled water. Add 10 drops each of NaCN and hydroxylamine hydrochloride solutions. Then add 2.5 mL NaOH solution and 10-15 drops of Calcon indicator. Pink colour will appear. Now in another China dish, prepare a blank solution exactly in the same manner, taking 5 mL of distilled water instead of standard Calcium solution. The blank solution will be blue in colour, but if not, it will turn blue after the addition of 2-3 drops of EDTA solution. Keep this blue blank along side for comparing the end point. Now take the earlier China dish containing the pink coloured standard calcium solution and place it on the magnetic stirrer plate. Put a glass coated iron needle in it and start stirring. Titrate the solution with EDTA until the colour of the solution is just changed from pink to blue matching the blank. Note volume of EDTA used. On the basis of calculations, $N_1V_1=N_2V_2$, necessary dilutions/concentrations are made so that the normality of the EDTA solution is exactly equal to 0.01N.
3. **Calcon indicator solution (0.5%):** Dissolve 0.5 g of Calcon reagent (Solo chrome dark blue and Erichrome blue black R) in 100 mL of 95% ethanol taken in 250 mL beaker. Store in a plastic bottle. Prepare fresh every 2 weeks.

4. **Sodium hydroxide solution (4N or 16%):** Weigh 160 g pure NaOH in a 600 mL beaker. Dissolve it by adding adequate distilled water and transfer it to 1 liter volumetric flask. Make the volume by distilled water and store the solution in a glass reagent bottle.
5. **Sodium cyanide solution (2%):** Accurately weigh 4 g NaCN in a 400 mL beaker. Add 200 mL distilled water and dissolve the salt. Store the solution in a glass bottle of optimum size. NaCN is a poison, so handle it with care.
6. **Hydroxylamine Hydrochloride solution (5%):** Weigh 5 g reagent hydroxylamine hydrochloride in a 250 mL beaker. Add 100 mL of distilled water to dissolve it. Store the solution in a plastic bottle. It should be prepared fresh every week.

Procedure

(i) In a 100 mL China dish, add 5 mL stock solution of plant digest (0.5g plant material in 50 mL aliquot/ solution) with pipette and dilute it to 10 mL with distilled water.

(ii) Add 10 drops each of 2% NaCN and hydroxylamine hydrochloride solutions.

(iii) Add 2.5 mL 4N NaOH solution and then add 10-15 drops of Calcon indicator. Pink to reddish colour will appear.

(iv) In another China dish, prepare a blank solution exactly in the same way, taking 5 mL of distilled water instead of the test solution (aliquot). In this case blue colour will appear, but if not, add few (2-3) drops of the EDTA solution to get the same. Keep this blue blank along side for comparing the end point of the test solution.

(v) Place the earlier China dish containing the pink coloured test solution on the magnetic stirrer top. Put a glass coated iron needle in it and start stirring.

(vi) Titrate the solution with 0.01N EDTA until the pink colour just changes to blue matching the blank. Note the volume of the EDTA used, say x mL. Calculate Ca in the aliquot and express the results in appropriate units.

Calculations

Weight of the plant material taken for the digestion	= 0.5 g
Final volume of the plant digest/aliquot made	= 50 mL
Therefore, dilution factor =50/0.5	=100 times
Volume of the aliquot taken for titration	= 5 mL
Volume of 0.01N EDTA used for titration	= x mL

Total Ca + mg (me/100 g plant) =

$$\frac{\text{mL of EDTA used X Normality of EDTA X dilution factor}}{\text{mL of aliquot taken for titration}} \text{ X } 100$$

$$= \frac{\text{x mL X 0.01N X 100 X 100}}{\text{5 mL}} = 20 \text{ x}$$

Or, total Ca in plant sample (%)

$$= \frac{\text{20 x X equivalent weight of Ca}}{1000}$$

$$= \frac{\text{20 x X 20}}{1000} = 0.4 \text{ x}$$

Interpretation of Results

Depending on the concentration of Ca determined in the plant material, a sample can be rated as sufficient or deficient.

Sufficient = 0.2 – 1.0 g Ca per 100 g plant material

Deficient < 0.2 g Ca per 100 g plant material

2.1.2. *Determination of Magnesium*

Determination of magnesium by titration by EDTA is done at pH 10 in the presence of Erichrome Black-t indicator. The end point of titration is more distinct than calcium alone, where, indicator colour change at end point is not so sharp. In titration at pH 10, it is not only Mg which is titrated but both Ca + Mg are titrated together and Mg is determined from the difference of Ca + Mg and Ca.

Principle

Magnesium in a plant digest (solution/aliquot) can be determined by titration of a known volume of the aliquot against the standard solution of EDTA using Erichrome Black-T as indicator at pH 10 in the presence of NH_4Cl-NH_4OH buffer. Adequate concentrations of NaCN and hydroxylamine hydrochloride are also used to remove the interference of the heavy metals. The end point of titration will be the change of colour from purple to blue, when whole of Mg gets complexed with EDTA. If Ca is also present in the aliquot/solution this titration will estimate both Ca and Mg, i.e., Ca+Mg.

Apparatus

Magnetic stirrer, glass coated iron needles, ceramic china dishes of 100 mL capacity, wash bottles, micro burette, pipettes, beakers, volumetric flasks, etc.

Reagents

1. **Standard solution of Magnesium (0.01N):** Weigh 0.1216 g unoxidized Mg metal and dissolve it in minimum volume (10-15 mL) of 1N HCl taken in a 600 mL beaker. Add nearly 400 mL of distilled water and transfer the contents of the beaker to 1 liter volumetric flask. Finally, make the volume to 1 litre. If, the magnesium metal is not available, MgSO4.7H2O can be used for making the required standard solution of magnesium by dissolving 1.2325 g of the salt in distilled water and then making the final volume to 1 litre.

2. **Standard solution of EDTA (0.01 N):** Prepare this solution as described under section 2.1.1, above and standardize it as under :

   Standardization of EDTA solution: Pipette 5 mL standard magnesium solution (0.01N) in a 100 mL China dish. Dilute it to 10 mL with distilled water. Add 10 drops each of NaCN and hydroxylamine hydrochloride solutions. Then add 10 mL of NH4OH- NH_4Cl buffer solution and 10 drops of Erichrome Black-T indicator. Purple colour will appear. Now, in another China dish, prepare a blank solution exactly in the same manner, taking 5 mL of distilled water instead of standard magnesium solution. The blank solution will be blue in colour, but if not, it will turn blue with the addition of 2-3 drops of EDTA solution. Keep this blue blank along side for comparing the end point of test solution. Now take the earlier China dish containing the purple coloured standard magnesium solution and place it on the magnetic stirrer plate. Put a glass coated iron needle in it and start stirring. Titrate the solution with EDTA until the colour of the solution is changed from purple to blue matching the blank. Note volume of EDTA used. On the basis of calculations, $N_1V_1=N_2V_2$, necessary dilutions are made so that the normality of the EDTA solution is exactly equal to 0.01N.

3. **Erichrome Black-T indicator solution (0.5%):** Dissolve 0.5 g Erichrome Black-T reagent in 100 mL 95% ethanol taken in 250 mL beaker and store it in a plastic bottle. Prepare fresh solution every 2 weeks.

4. **Ammonium chloride-Ammonium hydroxide buffer:** Take 67.5 g of NH4Cl in a 1 liter beaker and add 400 mL of distilled water to dissolve the salt. To this add 570 mL of 88% NH_4OH solution and make volume to 1 liter.

5. **Sodium cyanide solution (2%):** Prepare this solution as described under section 2.1.1, above.

6. **Hydroxylamine hydrochloride solution (5%):** Prepare this solution as described under section 2.1.1, above.

Procedure

(i) In a 100 mL China dish, add 5 mL stock solution of the plant digest (0.5g plant material in 50 mL aliquot/ solution) with pipette and dilute it to 10 mL with distilled water.

(ii) Add 10 drops each of 2% NaCN and hydroxylamine hydrochloride solutions.

(iii) Add 10 mL of NH_4OH- NH_4Cl buffer and 10 drops of Erichrome Black-T indicator. Purple colour will appear.

(iv) In another China dish, prepare a blank solution exactly in the same way, taking 5 mL of distilled water instead of the test solution (aliquot). In this case blue colour will appear, but if not, add few (2-3) drops of EDTA solution to get the same. Keep this blue blank along side for comparing the end point of the test solution.

(v) Place the earlier China dish containing the purple coloured test solution on the magnetic stirrer top. Put a glass coated iron needle in it and start stirring.

(vi) Titrate the solution with 0.01N EDTA until the purple colour changes to blue matching the blank. Titration near the end point should be done very slowly. Note the volume of the EDTA used, say y mL.

Calculations

Weight of the plant material taken for the digestion = 0.5 g

Final volume of the plant digest/aliquot made = 50 mL

Therefore, dilution factor =50/0.5=100 times

Volume of the aliquot taken for titration = 5 mL

Volume of 0.01N EDTA used for titration = y mL

Total ca + Mg (me per 100 g plant)

$$= \frac{\text{mL of EDTA used X Normality of EDTA X dilution factor X 100}}{\text{mL of aliquot taken for titration}}$$

$$= \frac{\text{y mL X 0.01N X 100 X 100}}{\text{5 mL}} = 20\text{ y}$$

As calculated already under section 2.1.1 above,

Total Ca (me per 100 g of plant material) = 20 x

Therefore, total Mg (me per 100 g of plant material) = (Ca+Mg) – Ca

$$= 20y - 20x$$

$$= 20\,(y - x)$$

Or, total Mg in plant sample (%)

$$= \frac{20\,(y - x) \times \text{Equivalent weight of Mg}}{1000}$$

$$= \frac{20\,(y - x) \times 12}{1000} = 0.24\,(y - x)$$

(c) Interpretation of Results

Depending on the concentration of Mg determined in the plant material, a sample can be rated as sufficient or deficient :

Sufficient = 0.1- 0.4 g Mg per 100 g plant material

Deficient < 0.1 g Mg per 100 g plant material

2.2. Determination of Ca and Mg by Emission Spectrophotometry

In the flame emission spectrophotometry technique, the quantity of emission (radiation energy) produced by the element to be determined is compared with its known quantity in the solution. As compared to titrimetric and gravimetric method, it is rapid and sensitive method to determine K, Na, Ca, Mg and other ions with accuracy of 1-3% (Jackson, 1967). This method saves lot of time, when large number of soil and plant samples are to be analyzed .

Principle

The emission spectrometry/flame photometry is based on the basic principle that when a fine mist of solution containing various atoms or ions like Na, K, Ca, Mg, etc. , is sprayed over a flame with the help of an atomizer, the atoms or ions absorb heat energy and get excited. They emit energy in the form of characteristic electromagnetic radiations, when they come back to the ground state. In a flame photometer there is a provision to isolate or filter the characteristic radiations of a particular atom/element through specific optical filters. The energy emitted will be directly proportional to the number of atoms excited which will depend on their concentration in the test solution. The energy so emitted is made to fall on the photocell of the flame photometer, which converts these radiations into electric energy. This energy is amplified and read on the galvanometer of the flame photometer.

2.2.1. *Determination of calcium*

Apparatus

Emission spectrophotometer/Flame photometer, volumetric flasks (50 mL, 100 mL, 500 mL, 1000 mL.), beakers, pipettes, etc.

Reagents

1. **1000 ppm stock solution of Ca:** Weigh 2.4973 g of $CaCO_3$ and transfer it into a 600mL beaker. Add to it 10 mL of 1N HCl and boil to expel CO_2. Allow it to cool and then add about 500 mL distilled water. Transfer the contents in a 1 litre volumetric flask and make the final volume.
2. **0.4N HCl base electrolyte solution:** It is prepared by the dilution of the concentrated HCl (11.6 N). This is used to prepare standard solutions of Ca from the 1000 ppm stock solution for preparing standard curve.

Procedure

A. Preparation of standard curve for Ca

(a) Preparation of standard solutions of Ca

Prepare 50 ppm solution of Ca by pipetting 25 mL of the 1000 ppm stock solution in a 500 mL volumetric flask and making the final volume with 0.4 N HCl base electrolyte solution. From the 50 ppm Ca solution, prepare a series of standard solutions of 0, 2.5, 5, 10, 15, 20, 30 and 40 ppm by taking 0, 5, 10, 20, 40, 60 and 80 mL of the solution in 100 mL volumetric flasks and making the final volume up to the mark with 0.4 N HCl base electrolyte solution.

(b) Calibration of flame photometer

(i) Switch on the flame photometer and apply Ca filter.

(ii) Feed distilled water to the instrument and set the zero of the galvanometer scale with the set zero knob.

(iii) Now feed the blank (0.4 N HCl base electrolyte) and reset the zero of the galvanometer.

(iv) Then feed the instrument with the highest concentration i.e., 50 ppm Ca solution and adjust the 100 of the scale with the amplifier knob.

(v) Repeat steps 3 and 4 to ensure proper calibration.

(c) Readings of standard solutions

Feed all the standard solutions (0, 2.5, 5, 10, 15, 20, 30, 40 and 50 ppm) one by one starting from the lowest concentration and record the galvanometer readings against each one. Repeat the above step once more for ensuring the consistency of the observations.

(d) Plotting of the curve

1. Take a simple graph paper and plot concentrations of the standard solutions of Ca on the x-axis.
2. On the y-axis of the graph paper, plot the readings of the instrument recorded against each concentration, i.e., 0, 2.5, 5, 10, 15, 20, 30, 40 and 50 ppm.
3. Draw a curve starting from the origin (zero) and joining all the points up to the highest point (100). This will give the standard curve for comparing the concentrations in the test solutions.

B. Analysis of test solution for Ca

(a) Observations

1. Calibrate the instrument by following all the steps for the calibration of the flame photometer as listed above [A.(b)].
2. Feed the test solution in which Ca concentration is to be determined and note down the reading.
3. Compare the reading of the test solution with the standard curve and note down the concentration against the reading, let it be x ppm.

(b) Calculations

Observed concentration of Ca in test solution	= x ppm
Weight of plant material taken for the digestion	= 0.5 g
Total volume of the digest/aliquot made	= 50 mL
Therefore, dilution factor = 50/0.5	= 100 times

So, total Ca in plant sample

= observed concentration X dilution factor

= x ppm X 100 times

= 100 x ppm.

Or, total Ca (%) in plant sample = 0.01 x

(c) Interpretation of Results

Depending on the concentration of Ca determined in the plant material, a sample can be rated as sufficient or deficient.

Sufficient = 0.2 – 1.0 g Ca per 100 g plant material

Deficient < 0.2 g Ca per 100 g plant material

2.2.2. *Determination of magnesium*

Apparatus

Emission spectrophotometer/Flame photometer, volumetric flasks (50 mL, 100 mL, 500 mL, 1000 mL.), beakers, pipettes, etc.

Reagents

1. **1000 ppm stock solution of Mg:** Dissolve 1.000 g of metallic Mg foil in 10 mL of 1N HCl and dilute it to 1 litre with distilled water. If metallic Mg foil is not available, $MgSO_4.7H_2O$ can be used. To make 1000 ppm solution of Mg, dissolve 10.1352 g of $MgSO_4.7H_2O$ salt in water and make up the final volume to 1 litre.
2. **0.4N HCl base electrolyte solution:** It is prepared by diluting concentrated HCl (11.6 N). It is used to prepare standard solutions of Mg from the 1000 ppm stock solution for preparing standard curve.

Procedure

A. Preparation of standard curve for Mg

(a) Preparation of standard solutions of Mg

1. Prepare 50 ppm solution of Mg by pipetting 25 mL of the 1000 ppm stock solution in a 500 mL volumetric flask and making the final volume with 0.4 N HCl base electrolyte solution.
2. From the 50 ppm Mg solution, prepare a series of standard solutions of 0, 2.5, 5, 10, 15, 20, 30 and 40 ppm by taking 0, 5, 10, 20, 30, 40, 60 and 80 mL of the solution in 100 mL volumetric flasks and making the final volume up to the mark with 0.4 N HCl base electrolyte solution.

(b) Calibration of flame photometer

(i) Switch on the flame photometer and apply Mg filter.

(ii) Feed distilled water to the instrument and set the zero of the galvanometer scale with the set zero knobs.

(iii) Now feed the blank (0.4 N HCl base electrolyte) and reset the zero of the galvanometer.

(iv) Then feed the instrument with the highest concentration i.e., 50 ppm Mg solution and adjust the 100 of the scale with the amplifier knob.

(v) Repeat steps 3 and 4 to assure proper calibration.

(c) Readings of standard solutions

1. Feed all the standard solutions (0, 2.5, 5, 10, 15, 20, 30, 40 and 50 ppm) one by one starting from the lowest concentration and record the galvanometer readings against each one.
2. Repeat the above step once more for ensuring the consistency of the observations.

(d) Plotting of the curve

1. Take a simple graph paper and plot concentrations of the standard solutions of Mg on the x-axis.
2. On the y-axis of the graph paper, plot the readings of the instrument recorded against each concentration.
3. Draw a curve starting from the origin (zero) and joining all the points up to the highest point (100). This will give the standard curve to be used for comparing the concentrations in the test solutions.

B. Analysis of test solution

(a) Observations

1. Calibrate the instrument by following all the steps for the calibration of the flame photometer as listed above under section A. (b).
2. Feed the test solution in which Mg concentration is to be determined and note down the reading.
3. Compare the reading of the test solution with the standard curve and note down the concentration against the reading, let it be y ppm.

(b) Calculations

Observed concentration of Mg in test solution= y ppm

Weight of plant material taken for the digestion = 0.5 g

Total volume of the digest/aliquot made = 50 mL

Therefore, dilution factor = 50/0.5 = 100 times

So, total Mg in plant sample

= observed concentration X dilution factor

= y ppm X 100 times

=100 y ppm.

Or, total Mg (%) in plant sample = 0.01 y

(c) Interpretation of Results

Depending on the concentration of Mg determined in the plant material, a sample can be rated as sufficient or deficient.

Sufficient = 0.1- 0.4 g Mg per 100 g plant material

Deficient < 0.1 g Mg per 100 g plant material

3. REFERENCES

Hesse, P.R. (1971). *A Textbook of Soil Chemical Analysis.* John Murray Publishers, London.

Jackson, M.L. (1967). *Soil Chemical Analysis.* Prentice- Hall of India, New Delhi.

Jones, J.B., Wolf, B. and Mills, H.A. (1991). *Plant Analysis Handbook: A Practical sampling, Preparation, Analysis and Interpretation Guide.* Micro-Macro Publishing Inc. Athens, Georgia.

Kensworthy, A.L. (1964). *Fruit, Nut and Plantation Crops- A Guide for Collection of Foliar Samples for Nutrient Element Analysis.* Mimeographed. Horticulture Department, Michigan state University, East Lansing, USA. Pp. 39.

Tandon, H.L.S. (1984). *Methods of Analysis of Soil, Plant, Water and Fertilizers.* Fertilizer Development and Consultation Organization, New Delhi.

Research Methods in Plant Sciences : Allelopathy.
Volume 4. Plant Analysis
Eds. S.S. Narwal, O.P. Sangwan and O.P. Dhankhar
Scientific Publishers (India), Jodhpur (2007) (pp. 119-124)

10

SULPHUR

Mohinder Singh, M.L. Dixit and J.S. Dhankhar

Department of Soil Science, CCS Haryana Agricultural University, Hisar – 125 004, INDIA

1. INTRODCTION

Sulphur (S) has assumed great significance among the plant nutrients as its deficiency in soils has escalated during the last few decades due to (i) increased crop requirement of S owing to higher crop yields (ii) increased use of S free fertilizers, (iii) reduced use of S containing pesticides and (iv) decreased evolution of SO_2 gas from industries etc. Now, S is the fourth important macronutrient after N, P and K. It is taken up by the plants from the soil in anionic form SO_4^{-2}. Its primary role in the plants is for (i) synthesis of amino acids (cysteine, cystine and methionine) and proteins, (ii) as constituent of vitamins (biotin and thiamin), iron-S proteins, thioredoxins, sulpholipids, sulphate esters, glucosinolates, polysaccharide sulphates and lipid sulphates, etc. and (iii) improve milling and baking quality of cereals, quality of tobacco, pungency in onions and garlic and nutritive quality of forages. The S requirement of oil seeds is highest (12 kg t^{-1}) followed by pulses (8 kg t^{-1}) and cereals (3-4 kg t^{-1}).

The content of sulphur in plant parts gives an indication about the status of sulphur availability in soil as well as its deficiency and sufficiency levels at a given stage of plant growth. It is, therefore, imperative to estimate the concentration of this nutrient in plants of a given crop for arriving meaningful conclusions while planning the fertilizer recommendations.

2. METHODS TO DETERMINE SULPHUR

The determination of total S in plants involves two steps i.e. (a) digestion of plant sample by suitable reagents to convert total S into anionic sulphate form in solution and (b) estimation of sulphate S in the solution. The methods for dry ashing and wet digestion of plant samples are described in Chapter 4 of this book.

After conversion of total S in plants to anionic sulphate form in the digest either by dry ashing or by wet digestion, the sulphate S may be reduced using special apparatus [Johnson and Nishita (1952) and Johnson and Ulrich (1959)] and its colorimetric estimation as methylene blue. However, the reduction of sulphate S and its determination is tedious. The sulphate S can be determined by turbidimetric method (Chesnin and Yien, 1950 and Tabatabai and Bremner, 1970), which is simple and rapid.

2.1 Turbidimetric method

This method to determine sulphur in plants is commonly used in laboratories for its simplicity, speed and reasonable accuracy. It is based on the principle of spectrophotometry.

Principle

In the turbidimetric method, sulphate S in plant is precipitated to barium sulphate by barium chloride crystals. The intensity of the barium sulphate turbidity is measured on spectrophotometer at 420 nm wave length.

Apparatus and Equipment

Volumetric flasks (50 and 100 ml), Pipettes, Spectrophotometer, 30 and 60 mesh sieves

Reagents

(i) Barium chloride crystals ($BaCl_2$ $2H_2O$): Ground to pass 30 mesh sieve but retained on 60 mesh sieve

(ii) Gum acacia solution (0.25%)

(iii) Potassium sulphate (K_2SO_4)

Procedure

(i) Take 2 to 10 ml aliquot of plant digest depending upon the S content in a 50 ml volumetric flask.

(ii) Add distilled water to the flask to make 15 to 20 ml contents in flask.

(iii) Add 1.0g of barium chloride (30-60 mesh) crystals in the flask and shake vigorously for 1 to 2 minutes.

(iv) Add 1 to 2 ml of 0.25% gum acacia solution in the flask to keep the suspension in the colloidal form. Again shake for 1-2 minutes and make up the volume to 50 ml with distilled water. White turbidity of barlum sulphate is developed, which remains stable for 5-30 min.

(v) The intensity of the turbidity is measured on spectrophotometer at 420 nm as % transmittance. It is converted to μg S ml^{-1} with the help of the standard curve of S, prepared afresh on the same spectrophotometer.

Preparation of standard curve

(i) Prepare a working standard solution of 100 μg S ml^{-1} by dissolving 0.5434 g dried reagent grade K_2SO_4 in distilled water and dilute to one litre.

(ii) Pipette out 0, 1, 2, 5, 8, 10, 15 and 20 ml solution from the working solution of 100 μg S ml^{-1} in a 100 ml volumetric flask. Add distilled water to these flasks to make approximately 50 ml content.

(iii) Add 1.0 g barium chloride (30 to 60 mesh) crystals to the flask and shake for one min.

(iv) Add 2 ml 0.25% gum acacia solution to each flask and again shake for 1-2 min.

(v) Make up the volume to 100 ml by adding distilled water and again shake for 1-2 min. It will give 0, 1, 2, 5, 8, 10, 15 and 20 μg S ml^{-1} solution. The turbidity of barium sulphate will be developed.

(vi) Read the intensity of turbidity of the solution of above flasks at 420 nm on spectrophotometer as % transmittance.

(vii) Draw standard curve between concentration of S and % transmittance on a semi log paper and calculate the concentration of S in the aliquot taken in the flask.

Calculation of sulphur in plants

Weight of plant sample taken (W)	= 1g
Total volume of plant digest mode (V)	= 100 ml
Volume of aliquot taken from plant digest (V_1)	= Say 5 ml
Total volume of aliquot made (V_2) (for reading % Transmittance)	= 50 ml

$$\text{Dilution factor (DF)} = \frac{V}{W} \times \frac{V_2}{V_1} \times \frac{100}{1} \times \frac{50}{5} = 1000 \text{ times}$$

Let the concentration of S in (V_2) aliquot from standard curve be = Cμg Sml^{-1}

Total S in plant sample = (C X DF) = 1000Cμg Sg^{-1}

Alternatively, the total S content in the plant sample can be calculated as under :

$$\text{Total S in plant} = \frac{\mu g\ S\ ml^{-1} \text{ in aliquot} \times 50\ ml}{\text{Volume of aliquot taken (ml)}} \times \frac{\text{Total volume of digest (ml)}}{\text{Wt. of plant sample taken (g)}}$$

$$= \frac{C \times 50}{5} \times \frac{100}{1} = 1000C\ \mu g\ Sg^{-1}$$

The S content (%)in plant sample as under :

$$= \frac{1000C \times 10^2}{10^6} \times \frac{C}{10} = 0.10C\ \%\ S$$

Interpretation

The sulphur concentration in plant samples may vary according to species, plant age and its part. Low concentration leads to deficiency symptoms of the S in new leaves, which reduces the growth, development and yield. Sulphur in excess amount is non toxic in nature, but in industrial areas the acid rains may damage the crop. Singh (1999) has reported the critical levels of S in different crops (Table 1).

Table 1. Critical levels of S in select crop plants (45-55 day old).

Crop	S concentration in dry matter (%)		
	Deficient	Moderate	Sufficient
Rice, wheat, maize, millets	<0.20	0.20-0.30	>0.30
Groundnut, mustard, soybean, cowpea, brinjal, French bean, cucumber	<0.25	0.25-0.40	>0.40
Sunflower, linseed	<0.35	0.35-0.55	>0.55
Horse gram, pea, chickpea	<0.45	0.45-0.75	>0.75
Potato, cauliflower, spinach	<0.40	0.40-0.75	>0.75

Precautions

(i) The standard curve for each lot should be prepared afresh and separately.

(ii) While preparing the standard curve for S all the reagents and steps used in methods of plant digest should be followed.

(iii) Avoid contamination from chemicals, water, rubber tubing and stoppers, filter papers and glasswares etc, which may have S as impurities.

(iv) Always use standard analytical chemicals and reagents which are free from sulphur.

3. REFERENCES

Chesnin, L. and Yien, C.H. (1950) Turbidimetric determination of available sulphate. *Soil Science Society of America Proceedings* **15**: 149-151.

Johnson, C.M. and Nishita, H. (1952). Microestimation of sulfur in plant materials, soils and irrigation water. *Analytical Chemist* **24**: 736-742.

Johnson, C.M. and Ulrich, A. (1959). *Analytical Methods for Use in Plant Analysis*. California Agricultural Experimental Station Bulletin NO. 766.

Singh, M.V. (1999). *Sulphur Management of Oilseed and Pulse Crops*. Bulletin No. 3. Indian Institute of Soil Science, Bhopal, India. pp. 54.

Tabatabai, M.A. and Bremner, J.M. (1970). Simple turbidimetric method for determination of total sulfur in plant material. *Agronomy Journal* **62**: 805-806.

Research Methods in Plant Sciences : Allelopathy.
Volume 4. Plant Analysis
Eds. S.S. Narwal, O.P. Sangwan and O.P. Dhankhar
Scientific Publishers (India), Jodhpur (2007) (pp. 125-132)

11

MICRO-NUTRIENTS

R.S. Antil, R.P. Narwal and O.P. Sangwan

Department of Soil Science, CCS Haryana Agricultural University, Hisar 125 004

1. INTRODUCTION

Micronutrients were first recognized as a limiting factor in crop production in 1920 in Florida, USA. For the normal growth and development, crop plants require 18 elements, of these nine (Zn, Mn, Fe, Cu, B, Mo, Co, Nl and Cl) essential elements, are required in micro amounts, hence called micronutrients. They play important roles in plant metabolism as they form stable complexes with naturally occurring ligands. Although, they are required by the plant in very small quantities but they play major role in enzyme activation. The plant samples may be collected, processed and digested for chemical analysis as per chapter 3 and 4 in this book.

Generally, Atomic Absorption Spectrophotometer (AAS) is widely used to determine different micronutrient elements in plant samples. The use of instrument will depend on the purpose of analysis. If one element is to be estimated, then AAS is the best. But if more than one element is to be estimated in the same sample then Inductively Coupled Plasma Atomic Emission Spectroscopy (ICP-AES) can be used. With these instruments direct reading of the element/nutrient concentration in the digest can be taken and if any dilution of the aliquot is needed, the same is done and the dilution factor is applied.

2. DETERMINATION OF MICRONUTRIENTS

Among the nine micronutrients (Fe, Cu, Zn, Mn, B, Mo, Cl, Co and Ni), Fe, Cu, Zn, Mn, Co and Ni are heavy metals. These heavy metal elements are suitably estimated on Atomic Absorption Spectrophotometer (AAS), because their atoms do not get excited under ordinary flames, hence, cannot be estimated correctly by Emission Spectrophotometer. Whereas, B and Mo are determined separately with spectrophoto-meter (Colorimeter). Besides, Cl is usually not tested as it is not deficient in soil or plant. Ni when exceeds the permissible limit it becomes toxic to the plant, hence has been included in the list of toxic elements. However, its determination is given here and under toxic metals in chapter 13.

2.1. Determination of Zn, Mn, Cu, Fe, Ni and Co

Principle

The six micronutrients (Zn, Mn, Fe, Cu, Ni and Co) can be determined in acid digest of plant samples, with the help of AAS (Atomic Absorption Spectrophotometer). The AAS, is based on the principle that atoms of metallic elements (Zn, Mn, Fe, Cu, Ni, Co), which normally remain in ground state under flame conditions, absorb energy when subjected to radiation of specific wavelength. The absorption of radiation is proportional to the concentration of atoms of that element. The absorption of radiation by the atoms is independent of the wavelength of absorption and temperature of the flame.

Apparatus

Conical flasks, volumetric flasks, hot plate, pipettes, balance, wash bottle, Whatman No. 42 filter paper, AAS.

Reagents

Zinc Sulphate ($ZnSO_4 .7H_2O$), Potassium Per manganete ($KMnO_4$), Copper sulphate ($CuSO_4.5H_2O$), Ferrous Ammonium Sulphate $(NH_4)_2 SO_4 .FeSO_4 . 6H_2O$, Cobaltus cloride ($CoCl_2$) and Concentrated Sulphuric Acid (H_2SO_4) and Oxalic acid ($C_2H_2O_4 . 2H_2O$).

Preparation of standard curves

2.1.1. *Preparation of standard solutions of reagents*

i **Zinc standard solution**: Dissolve 0.439 g zinc sulphate ($ZnSO_4.7H_2O$) of AR grade in 200 mL of double distilled water in a 500 mL beaker. Add 5 mL of dilute H_2SO_4 (1:5) and make up the volume to 1.0 litre. This is a standard solution of 100 ppm

Zn. Dilute 10 mL of this standard solution to 100 mL in a volumetric flask to prepare stock solution of 10 ppm. Dilute 1,2,4 and 6 mL of stock solution (10 ppm) in a 100 mL volumetric flask to get working standards (0,0.1, 0.2, 0.4 and 0.6 ppm Zn).

ii **Manganese Standard Solution**: Dissolve 0.288 g potassium permanganate ($KMnO_4$) AR grade in 300 mL of double distilled water in a 500 mL beaker. Add 20 mL concentrated H_2SO_4, warm to about 60^{0}C and add oxalic acid solution drop by drop to make the solution colourless. Cool and transfer to a 1.0 litre measuring flask and make volume upto the mark. This is 100 ppm stock solution of Mn. Make use of this solution for preparing working standards (0, 0.2, 0.4, 0.8, 1, 2, 5 and 10 ppm).

iii **Copper Standard Solution**: Dissolve 0.392 g copper sulphate ($CuSO_4.5H_2O$) of AR grade in 300 mL of double distilled water and make up volume to 1.0 litre. This is standard solution of 100 ppm Cu. To prepare working standards, dilute 100 ppm Cu stock solution to obtain 1,2,4,6 and 8 ppm Cu.

iv **Iron Standard Solution:** Dissolve 0.702 g of AR grade ammonium ferrous sulphate $(NH_4)_2SO_4$ $FeSO_4.6H_2O$ in 300 mL of double distilled water in a 500 mL beaker. Add 5 mL of dilute H_2SO_4 (1:5) and transfer to a1 litre measuring flask and make volume up to the mark. This is a standard solution of 100 ppm Fe. Finally prepare working standards containing 1, 2, 4 and 6 ppm Fe by diluting appropriate volumes of 100 ppm Fe.

v **Cobalt Standard Solution :** Dissolve 0.220 g cobaltus chloride ($CoCl_2$) g AR grade in about 300 ml double distilled water taken in a 500 ml beaker. Dissolve the salt and transfer to a 1.0 litre volumetric flask and make up the volume upto the mark. This will give 100 ppm solution of Co. From this solution prepare working standard of 1, 2, 4, 6 and 8 ppm.

vi **Nickel Standard Solution**: Dissolve 1.0 g of Ni metal in 50 mL of HNO_3 (1 : 1 dilution). After the dissolution of Ni dilute the solution to 1.0 litre with double distilled water. This is a standard solution of 1000 ppm of Ni. Make use of this solution for the preparation of working standards (0, 0.2, 0.4, 0.8, 1.0, 2, 4, 8 and 10 ppm).

2.1.2. *Plotting of standard curves*

Feed the working standard solutions of each nutrient element to AAS and take the readings against their concentrations and then prepare the standard curves for each element by plotting AAS readings against Zn, Mn, Cu, Fe and Co concentrations.

Procedure

(i) First, apply the respective cathode lamp for the particular nutrient being determined and then feed the blank of the working standards to the AAS.

(ii) Feed the aliquots (plant digest containing 0.5 g plant material in 50 ml volume) directly or after a required dilutions if needed.

(iii) Note down the readings (and dilution factor if dilution was done) on the AAS. Let the reading be 'M'.

(iv) Compare the readings of each element with their standard curves prepared already, and calculate concentration of each nutrient element in ppm.

Calculations

Weight of plant material taken = 0.5 g

Volume made after acid digestion = 50 mL

Dilution factor = 50/0.5 = 100 times

Concentration of Zn, Mn, Fe, Cu, Ni or Co (ppm) in plant sample = M x 100

Where M stands for the Zn, Mn, Fe, Cu, Ni or Co concentration in aliquot as read from standard curve against the sample readings

Precautions

(i) Always grind plant samples in stainless steel grinding machine.

(ii) Use only double distilled water, corning/borosil glass apparatus and AR grade chemicals.

5.2. Determination of Boron (B)

Boron in the acid digest of plant sample can be determined by a spectrophotometer using azomethine-H reagent (Berger and Truog, 1939; Sippola and Ervco, 1977).

Equipment and apparatus

Spectrophotometer, a dispenser, boron-free glassware, viz., conical flasks, volumetric flasks, pipette.

Reagents

(i) **Concentrated HNO_3 and $HClO_4$**

(ii) **Azomethine-H reagent**: Dissolve 0.45 g of azomethine-H in 100 mL of 1% L-ascorbic acid solution in polypropylene reagent bottle. Fresh reagent should be prepared each week and stored in a refrigerator.

(iii) **Buffer solution**: Dissolve 250 g of ammonium acetate and 15 g of EDTA-disodium salt in 400 mL of distilled water and slowly add 125 mL of glacial acetic acid and mix.

Preparation of standard solutions

Dissolve 0.570 g boric acid (H_3BO_3) in distilled water and dilute to one litre to obtain a stock solution of 100 ppm B. Dilute 5 mL of stock solution to 100 mL with distilled water which will give a B concentration of 5 ppm. Pipette 0, 0.25, 0.50, 1.0, 2.0 and 4.0 mL of 5 ppm solution to a 25 mL volumetric flask to have working standards of 0, 0.05, 0.10, 0.20, 0.40 and 0.80 ppm B, respectively.

Preparation of standard curve

Read the intensity of the color of the solution on a spectrophotometer at 420 nm. Draw a standard curve by plotting the % transmittance (Y-axis) against B concentrations (X-axis) on a semi log paper.

Procedure

(i) Pipette 1 mL aliquot from the acid digested solution in a 25 mL volumetric flask and add 4 mL of buffer solution and mix well.

(ii) Add 4 mL of azomethine-H reagent and mix the contents.

(iii) The colour is allowed to develop for 1.0 h, and the volume is made up to the mark.

(iv) Intensity of the colour is measured on a spectrophotometer at 420 nm.

(v) Calculate the concentration of B against the reading from the standard curve as under.

Calculations

Weight of the plant material taken = 0.5 g

Volume made after acid digestion = V mL

Volume of aliquot taken = V_1 mL

Final volume made = 25 mL

Transmittance (%) as read from spectrophotometer = T

Concentration of B in the solution (ppm) against the transmittance T = C

First dilution = (V/0.5) times

Second dilution = ($25/V_1$) times

Total dilution = (V/0.5) x (25/V_I) times=Z

Content of B (ppm) in plant sample = C x Z

5.3. Determination of Molybdenum (Mo)

Principle

Molybdenum in the plant samples can be determined by ashing and fusion of sample. A known volume of the filtrate, containing measured amount of plant material ashed and fused with anhydrous Na_2CO_3, is taken and Mo content is determined spectrophoto-meterically.

Equipment and Apparatus

Spectrophotometer, hot plate, refrigerator, water bath, muffle furnace, separating funnel, glass stoppered flasks, volumetric flasks, pipettes.

Reagents

- Anhydrous Na_2CO_3.
- **50% Potassium iodide solution:** Dissolve 50 g potassium iodide in 100 mL of double distilled water (DDW).
- **50% Ascorbic acid solution:** Dissolve 50 g ascorbic acid in 100 mL of DDW.
- **10% Tartaric acid solution:** Dissolve 10 g Tartaric acid in 100 mL of DDW.
- **10% Thiourea solution:** Dissolve 10g thiourea in 100 mL of DDW and filter. Always prepare a fresh just before use.
- **Dithiol solution (Toluene-3,4-dithiol):** One g of dithiol (51°C) is added in 100 mL of 10% NaOH solution. The contents are warmed up to 51°C with frequent stirring for 15 minutes on a hot plate. Then 1.8 mL of thioglycolic acid is added and stored in refrigerator.
- Isoamyl acetate solution.
- **Ferrous ammonium sulphate solution**: Dissolve 63 g of ferrous ammonium sulphate in distilled water, add 10 mL concen-trated H_2SO_4 and make the volume to 1.0 litre.
- Ethyl alcohol (Ethanol).

Preparation of Standard solutions of Molybolate

Dissolve 0.184 g of AR grade ammonium molybdate in 100 mL of 0.1 N NaOH, make it slightly acidic with dilute HCl and make volume to 1.0 litre to obtain a stock solution of 100 ppm. Dilute 1, 2,

3, 4, 5, 6, 8,10 mLs of 100 ppm standard Mo solution to 1000 mL with water in volumetric flask to obtain respective working standards of 0.1, 0.2, 0.3, 0.4, 0.5, 0.6, 0.8, 1.0 ppm of Mo.

Preparation of standard curve

Read the colour intensity of different working standard solutions of Mo on a spectrophotometer at 680 nm by using red filter. Draw a standard curve by plotting the % transmittance (Y-axis) against Mo concentrations (X-axis) on a semi log paper.

Procedure

(i) Transfer 5 g material (1 to 10 g) in a platinum dish.

(ii) Ash it in an electric furnace at 500° C. as per Chapter 4 in this book.

(iii) After ashing, fuse the ash with 2 g of anhydrous Na_2CO_3 at 1000°C in the electric furnace, ensuring complete contact of the entire ash with the flux.

(iv) Cool the dish and drop the cake in a 250 mL beaker by inverting the dish.

(v) If cake does not detach, place the dish in beaker and disintegrate the cake in 100 mL water containing 2% ethanol (v/v).

(vi) Measure the suspension in a cylinder and filter through Whatman No.42 filter paper in a 250 mL beaker.

(vii) Transfer 20 mL of the filtrate in 250 mL separatoring funnels and add 0.25 mL of the ferrous ammonium sulphate solution and 20 mL of double distilled water and shake the funnel vigorously, up and down.

(viii) Add excess of the KI solution and clear the liberated iodine by adding ascorbic acid drop by drop on continuous stirring/shaking.

(ix) Add one mL of tartaric acid and 2 mL of the thiourea solution and again shake vigoursly.

(x) Add 5 drops of the dithiol solution and allow the mixture to stand for 30 min.

(xi) Add 10 mL of the isoamyl acetate and separate out the contents (green color) in glass cuvettes to read the colour intensity (% T) at 680 nm on a spectrophotometer using red filter.

Calculations

Weight of the plant material taken, = 5 g

Volume made after acid digestion = 100 mL

Volume of aliquot taken for color development = 20 mL

$$\text{Mo content (ppm) in plant sample} = C \times \frac{100}{20} \times \frac{1}{5} = C$$

Where C stands for Mo concentration obtained from standard curve against the reading (%T).

Interpretation of plant test results for micronutrients

Table 1 shows the deficient and sufficient levels of all micronutrient elements, for interpretation of results of plant analysis.

Table 1. Micronutrients content in mature plant tissue for classification as deficient and sufficient.

	Micronutrients concentration (mg/kg dry matter)	
Micronutrient	Deficient	Sufficient
Zn	<20	25-150
Fe	<50	50-250
Mn	<20	20-500
Cu	<4	5-20
B	<15	20-100
Mo	<0.01	0.5-0.1
Co	<10	10-20
Ni	<0.1	0.1-10

Source: Mortvedt *et al.* (1972), Tandon (1984), Jones *et al.* (1991) and Bhargav and Raghupati (1993).

6. REFERENCES

Berger, K.C. and Truog, E. (1939). Boron determination in soils and plants. *Industrial and Engineering Chemistry Analysis* **11**:540-45.

Bhargava, B.S. and Raghupati, H.B. (1993). Analysis of plant materials for macro and micronutrients. Pp. 49-82 In: *Methods of Analysis of Soils, Plants, Waters and Fertilizers.* (Ed. H.L.S. Tandon) FDCO, New Delhi.

Jones, J.B.Jr., Wolf, B. and Mills, H.A. (1991). *Plant Analysis Handbook: A Practical Sampling, Preparation, Analysis, and Interpretation Guide.* Micro-macro Publishing Inc. Athens, Georgia, USA.

Mortvedt, J.J, Giordand, P.M. and Lindsay, W.L. (1972). *Micronutrients in Agriculture.* Soil science society of America, Inc. Madison, Wisconsin, USA.

Sippola, J. and Ervco, R. (1977). Determination of boron in soils and plants by azomethine-II method. *Finnish Chemistry Letters* 138-40.

Tandon, H.L.S. (1984). *Methods of Analysis of Soils, Plants, Waters and Fertilizers.* Fertilizer Development and Consultation Organization, New Delhi.

Research Methods in Plant Sciences : Allelopathy.
Volume 4. Plant Analysis
Eds. S.S. Narwal, O.P. Sangwan and O.P. Dhankhar
Scientific Publishers (India), Jodhpur (2007) (pp. 133-136)

12

TOXIC METALS

R.P. Narwal, R.S. Antil, Balwan Singh,
A.V. Shanwal and D.S. Dahiya

Department of Soil Science, CCS Haryana Agricultural University, Hisar 125 004

1. INTRODUCTION

Recently, public awareness about metals (cadmium, lead, nickel and chromium) contamination in environment has increased. The disposal of sewage/ industrial effluents and city wastes on agricultural land is becoming a major problem. Such materials contain various toxic metals (cadmium, lead, nickel, and chromium) and pollute the soils. Their accumulation/high concentration in soils, causes toxicity (plants, organisms, animals and humans) and adversely affect the nutrient transformations, microbial population and enzyme activities in plants.

2. ANALYTICAL METHODS

The plant samples may be collected, processed and digested for chemical analysis as per Chapters 3 and 4 of this book. Many classical colorimetric procedures have been described to determine the toxic metals in plant samples. However, presently atomic absorption spectrophotometer (AAS) or Inductively Coupled Plasma

Atomic Emission Spectroscopy (ICP-AES) methods are used. With these instruments, direct reading of the digest can be taken.

3. DETERMINATION

Generally, Atomic Absorption Spectrophotometer is used to determine various toxic metals in plant samples. If more that one elements are to be estimated in plants at one time, ICP-AES technique is used.

Principle

The AAS is based on the principle that atoms of metallic elements (Cd, Ni, Pb and Cr), which normally remain in ground state, absorb energy when subjected to radiation of specific wavelength. The absorption of radiation is proportional to the concentration of atoms of that element. The absorption of radiation by the atoms is independent of the wavelength of absorption and temperature of the flame.

Apparatus

Conical flasks, volumetric flasks, hot plate, pipettes, balance, wash bottle, Whatman No. 42 filter paper, AAS.

Reagents

Ni metal, Concentrated Nitric Acid (HNO_3), Cd metal, concentrated Hydrochloric Acid (HCl), Potassium Dichromate ($K_2Cr_2O_7$) and Pb metal

3.1. Preparation of standard solutions

(i) **Nickel Standard Solution**: Dissolve 1.0 g of Ni metal in 50 mL of HNO_3 (1 : 1 dilution). After the dissolution of Ni, dilute the solution to 1.0 litre with double distilled water. This is a standard solution of 1000 ppm of Ni. Use this solution to prepare working standards (0, 0.2, 0.4, 0.8, 1.0, 2, 4, 8 and 10 ppm).

(ii) **Cadmium Standard Solution**: Dissolve 1.0 g of Cd metal in 50 mL of HCl (1 : 1 dilution). Then dilute to 1.0 litre with double distilled water. This is a standard solution of 1000 ppm. Use this solution to prepare working standards (0, 0.05, 0.1, 0.2, 0.4, 0.6, 0.8, 1.0 and 2.0 ppm).

(iii) **Chromium Standard Solution**: Dissolve 2.8282 g of reagent - grade potassium dichromate ($K_2Cr_2O_7$) in water, adjust to pH 6 to 7, and dilute to 1.0 litre with double distilled water. This is a standard solution of 1000 ppm. Use this solution to prepare working standards (0, 0.5, 1.0, 2.0, 4, 6, 8 and 10 ppm).

(iv) **Lead Standard Solution**: Dissolve 0.100 g of 99.9% Pb metal in 1:1 nitric acid (HNO_3). Dilute to 1.0 litre to give 100 ppm Pb solution. Finally prepare working standards containing 1, 2, 4, 6, 8 and 10 ppm Pb, by diluting 100 ppm Pb stock solution.

3.2. Preparation of standard curves

Feed the working standard solutions of each metal to AAS and record the readings against their concentrations. Prepare the standard curves by plotting AAS readings against Ni, Cd, Cr and Pb concentrations.

Procedure

1. First, apply the respective cathode lamp for the particular nutrient being determined and then feed the blank of the working standards to AAS.
2. Feed the aliquots (plant digest containing 0.5 g plant material in 50 ml volume) directly or after a required dilutions if needed.
3. Note down the readings (and dilution factor if dilution was done) on the AAS. Let the reading be 'M'.
4. Compare the readings of each element with their standard curves prepared already, and calculate concentration of each nutrient element in ppm.

Calculations

Weight of plant material taken = 0.5 g

Volume made after acid digestion = 50 mL

Dilution factor = 50/0.5 = 100 times

Concentration of Cd, Ni, Pb or Cr (ppm) in plant sample = M x 100

Where, M stands for the Cd, Ni, Pb or Cr concentration in aliquot as read from standard curve against the sample readings

Precautions

(i) Always grind plant samples in stainless steel grinding machine.

(ii) Use only double distilled water, corning/borosil glass apparatus and AR grade chemicals.

Interpretation of plant test results for toxic metals

Table 1 shows the deficient and sufficient levels of all toxic elements, namely Cd, Cr, Pb and Ni for necessary interpretation of the results of plant analysis.

Table 1. Critical concentration (mg kg^{-1} dry matter) of various toxic metals in plants and in diet of animals. Higher levels are toxic.

Nutrients	Plant	Animals
Cd	5-10	0.5-1.0
Cr	1-2	50-3000
Pb	10-20	10-30
Ni	20-30	50-60

Source: Extracted and modified from Mortvedt *et al.* (1972), Sauerbeck (1982).

6. REFERENCES

Mortvedt, J.J, Giordano, P.M. and Lindsay, W.L. (1972). *Micronutrients in Agriculture*. Soil Science Society of America, Madison, Wisconsin, USA.

Sauerbeck, K.A. (1982). Which heavy metal concentrations in plant should not be exceeded to avoid detrimental effects on their growth. *Landw. Forsch.* Sonderh. **39**: 108-129.

SUBJECT INDEX

AUTHOR INDEX